Sustainable Civil Infrastructures

About this Series

Sustainable Infrastructure impacts our well-being and day-to-day lives. The infrastructures we are building today will shape our lives tomorrow. The complex and diverse nature of the impacts due to weather extremes on transportation and civil infrastructures can be seen in our roadways, bridges, and buildings. Extreme summer temperatures, droughts, flash floods, and rising numbers of freeze-thaw cycles pose challenges for civil infrastructure and can endanger public safety. We constantly hear how civil infrastructures need constant attention, preservation, and upgrading. Such improvements and developments would obviously benefit from our desired book series that provide sustainable engineering materials and designs. The economic impact is huge and much research has been conducted worldwide. The future holds many opportunities, not only for researchers in a given country, but also for the worldwide field engineers who apply and implement these technologies. We believe that no approach can succeed if it does not unite the efforts of various engineering disciplines from all over the world under one umbrella to offer a beacon of modern solutions to the global infrastructure. Experts from the various engineering disciplines around the globe will participate in this series, including: Geotechnical, Geological, Geoscience, Petroleum, Structural, Transportation, Bridge, Infrastructure, Energy, Architectural, Chemical and Materials, and other related Engineering disciplines.

More information about this series at http://www.springer.com/series/15140

John Calautit · Fernanda Rodrigues
Hassam Chaudhry · Haşim Altan
Editors

Towards Sustainable Cities in Asia and the Middle East

Proceedings of the 1st GeoMEast International Congress and Exhibition, Egypt 2017 on Sustainable Civil Infrastructures

Springer

2017 GEOMEAST

SSIGE
SOIL-STRUCTURE INTERACTION GROUP IN EGYPT
المجموعة المصرية للهندسة الجيوتقنية والبنية التحتية

Editors
John Calautit
University of Nottingham
Nottingham
UK

Hassam Chaudhry
Heriot-Watt University Dubai
Dubai
United Arab Emirates

Fernanda Rodrigues
University of Aveiro
Aveiro
Portugal

Haşim Altan
University of Sharjah, UAE
Sharjah
United Arab Emirates

ISSN 2366-3405 ISSN 2366-3413 (electronic)
Sustainable Civil Infrastructures
ISBN 978-3-319-61644-5 ISBN 978-3-319-61645-2 (eBook)
DOI 10.1007/978-3-319-61645-2

Library of Congress Control Number: 2017946438

Printed on acid-free paper

This Springer imprint is published by Springer Nature
The registered company is Springer International Publishing AG
The registered company address is: Gewerbestrasse 11, 6330 Cham, Switzerland

Preface

Toward building sustainable and longer civil infrastructures, the engineering community around the globe continues undertaking research and development to improve existing design, modeling, and analytical capability. Such initiatives are also the core mission of the Soil-Structure Interaction Group in Egypt (SSIGE) to contribute to the ongoing research toward sustainable infrastructure. This conference series "GeoMEast International Congress and Exhibition" is one of these initiatives.

Ancient peoples built their structures to withstand the test of time. If we think in the same way, our current projects will be a heritage for future generations. In this context, an urgent need has quickly motivated the SSIGE and its friends around the globe to start a new congress series that can bring together researchers and practitioners to pursue "Sustainable Civil Infrastructures." The GeoMEast 2017 is a unique forum in the Middle East and Africa that transfers from the innovation in research into the practical wisdom to serve directly the practitioners of the industry.

More than eight hundred abstracts were received for the first edition of this conference series "GeoMEast 2017" in response to the Call for Papers. The abstracts were reviewed by the Organizing and Scientific Committees. All papers were reviewed following the same procedure and at the same technical standards of practice of the TRB, ASCE, ICE, ISSMGE, IGS, IAEG, DFI, ISAP, ISCP, ITA, ISHMII, PDCA, IUGS, ICC, and other professional organizations who have supported the technical program of the GeoMEast 2017. All papers received a minimum of two full reviews coordinated by various track chairs and supervised by the volumes editors through the Editorial Manager of the SUCI "Sustainable Civil Infrastructure" book series. As a result, 15 volumes have been formed of the final +320 accepted papers. The authors of the accepted papers have addressed all the comments of the reviewers to the satisfaction of the track chairs, the volumes editors, and the proceedings editor. It is hoped that readers of this proceedings of the GeoMEast 2017 will be stimulated and inspired by the wide range of papers written by a distinguished group of national and international authors.

Publication of this quality of technical papers would not have been possible without the dedication and professionalism of the anonymous papers reviewers. The names of these reviewers appear in the acknowledgment that follows. For any additional reviewers whose names were inadvertently missed, we offer our sincere apologies.

We are thankful to Dr. Hany Farouk Shehata, Dr. Nabil Khelifi, Dr. Khalid M. ElZahaby, Dr. Mohamed F. Shehata, and to all the distinguished volumes editors of the proceedings of the GeoMEast 2017. Appreciation is extended to the authors and track chairs for their significant contributions. Thanks are also extended to Springer for their coordination and enthusiastic support to this conference. The editors acknowledge the assistance of Ms. Janet Sterritt-Brunner, Mr. Arulmurugan Venkatasalam in the final production of the 15 edited volumes "Proceedings of GeoMEast 2017".

Contents

The Iron Triangle of Projects Management: Quality, Schedule and Cost of Road Infrastructure Projects in Egypt

Ahmed Abu El-Maaty[1(✉)], Ahmed Yousry Akal[2], and Saad A. El-Hamrawy[1]

[1] Highway and Airport Engineering, Faculty of Engineering, Menoufia University, Shibin El Kom, Egypt
maaty5000@yahoo.com

[2] Civil Engineering Department, Faculty of Engineering, Menoufia University, Shibin El Kom, Egypt
a_yousry85@yahoo.com

Abstract. One of the most widely used project performance measures is the "iron triangle" consisting of schedule, cost, and quality at the three vertices. The importance of executing project according to the expected quality level within the assigned budget and time constraints has been recognized since ancient times and it is considered as a significant performance measure of any construction project. Globally, various studies have revealed that road infrastructure contribute to the economic growth. Furthermore, for an optimal economic return, road infrastructure has to be delivered with reduced cost, on time, and in accordance with technical specifications.

Despite the importance of road projects and their impact on the national economy, little studies have been conducted in Egypt for enhancing the management of such projects. For that, it is of key importance to accomplish a study on road projects in Egypt to judge these projects' performance and determine whether these projects suffer from the issues related to quality, time, and, cost or not. This paper aims at finding out more knowledge about the attributes of the iron triangle (quality, time, and cost), using road projects in Egypt as a case study, through: (a) presenting definitions of quality, time overrun, and cost escalation of road projects, and (b) carrying out a field survey of the constructed road projects in Egypt to determine the percentages of quality level, time overrun, and cost escalation.

The study's results show that roads infrastructure contributes to the economic growth and is considered one of the most important industries affecting the Gross Domestic Product in Egypt. Furthermore, the results of the field survey which have been conducted on 56 road projects in Egypt present that the quality level percentage of the majority of road projects in Egypt ranges between 60% to 80% with average percentage = 71.76%. Furthermore, all the surveyed projects suffering from time overrun and cost escalation with average percentage = 73.80% and 46.30% respectively.

Keywords: Iron triangle of projects management · Quality · Time overrun · Cost escalation · Management of road projects and Egyptian economy

J. Calautit et al. (eds.), *Towards Sustainable Cities in Asia and the Middle East*,
Sustainable Civil Infrastructures, DOI 10.1007/978-3-319-61645-2_1

1 Introduction

Project success is basically to gain the project objectives that are classically defined by the need to complete a project on time, within the budget, and with appropriate quality (Abedi *et al.* 2011). Cost, time, and quality have their proven importance as the prime factors for project success. Furthermore, a project may not be regarded as a successful endeavor until it satisfies the cost, time, and quality limitations applied to it (Mahamid 2013). Over the last 50 years, quality, time, and cost have become inextricably linked with measuring the success of project management. This is perhaps not surprising, since over the same period these criteria have usually been included in the description of project management. These three factors remain in constant tension as can be seen in Fig. 1 (Atkinson 1999).

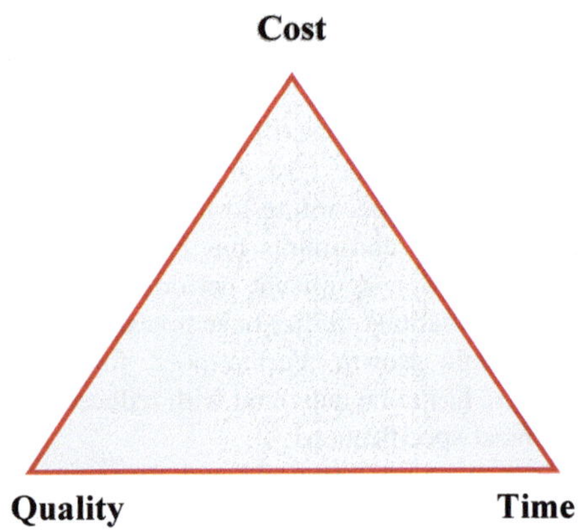

Fig. 1. The iron triangle of projects management (Atkinson 1999)

The construction industry is one of the largest if not the largest, major industries in any country in the world, whether developing or developed (Ozdemir 2010). Moreover, the construction industry is the tool through which a society achieves its goals of urban and rural development (Enshassi *et al.* 2006). Otherwise, it is one of the sectors that provide important ingredients for the development of an economy and its importance lies in the function of its products, which provide the foundation for industrial production and its impacts on the national economy cannot be measured by the value of its output or the number of persons employed in its activities alone (Alhomidan 2013; Hendrickson and Au 1989). The construction industry accounts 6–9% of the Gross Domestic Product (GDP) in many countries (Chitkara 2004). The construction industry in Egypt accounts for 4.7% of Egypt's GDP, making it one of the most important industries for the country's economic progress (Timetric 2012).

Despite, the importance of the construction industry in Egypt for the economic progress especially roads construction industry, a lot of the constructed projects are mismanaged, incurring much delay, cost escalation, and quality shortfall. Project management as defined by Oberlender (1993) is "*the art and science of coordinating people, equipment, materials, money, and schedules to complete a specified project on time, within approved cost, and according to the specified quality standards*". Project management tools and techniques play an important role in the effective management of a project. Therefore, a good project management lies in the management tools and techniques used to manage the project (Giridhar and Ramesh 1998).

A construction project such as road comprises two distinct phases: the preconstruction phase (the period between the initial conception of the project and signing of the contract) and the construction phase (the period after award of the contract when the actual construction is going on) (Giridhar and Ramesh 1998). During the construction phase several factors, however, are particularly noteworthy because of their significant impacts on the quality, time, and cost of construction (Falqi 2004). As a result, the construction projects may be exposed to delay in completion, cost overrun, and quality defects (Ibn-Homaid et al. 2011). The resultant poor performance in the delivery of construction projects in terms of cost, time, and quality has far reaching effects on the economy which, if left unchecked, could retard the national development (Kaliba 2010).

In order to overcome the negative impacts of delay, cost overrun, and quality defects on the performance of road projects in Egypt it is essential to improve the management of these projects through finding out more knowledge about the fundamental criteria for controlling and upgrading the projects performance (quality, time, and cost) by determining:

(a) Definitions of quality, time overrun, and cost escalation of road projects.
(b) Percentages of quality level, time overrun, and cost escalation through carrying out a field survey of the constructed road projects in Egypt.

2 The Need to Improve Management of Road Projects in Egypt

The wealth of any nation is gauged by its performance in infrastructure provision through its construction industry (Kaliba *et al.* 2009). Transportation infrastructure, specifically road infrastructure, is the backbone of economic and social development of countries. It is useful for communication and trade exchanges locally and internationally (Akoa 2011). Fan and Kang (2005) assert that roads contribute to economic growth and poverty reduction. Road infrastructure impacts on overall economic growth, agricultural growth, urban growth, urban poverty reduction, and rural poverty reduction. Without infrastructure, efficient markets, adequate health care, a diversified rural economy, and sustainable economic growth will remain elusive.

Fan and Kang (2005) continue to argue that transportation infrastructure is an effective factor of production. Power consumption and health conditions are positively correlated with the availability of road infrastructure. Furthermore, Nadiri and Mamuneas (1998) conclude that an increase in the stock of highway capital has an initial direct productivity effect on business: it reduces the total cost of producing a given level of output in almost all industries. Cost reductions permit products to be sold at lower prices and lower prices can be expected to lead to output growth. In addition to this, Kaliba *et al*. (2009) confirm that road construction constitutes a major component of the construction industry for developing economies, which means much of the national budget on infrastructure development is channeled to the road construction projects.

Egypt's road network comprises 91,173 km of roads, divided into 67,728 of main roads and 23,445 km of artery roads. Furthermore, the road network in Egypt carries out 85% of domestic freight and 60% of passenger movement, which makes it one of the most important industries affecting the GDP (Farag 2012; Timetric 2012). In an effort to improve the Egyptian economy, the Government had recognized the importance of the road network to the national economy and outlined US$5.46 billion in investments to improve the road transport system to facilitate the more efficient movement of goods (PWC's Staff Reporter 2013). However, it is unfortunate that, the constructed and the under construction road projects in Egypt are infamous for quality shortfall, time overrun, and cost escalation problems.

In road projects the lower level of quality leads to maintenance and repair works, which directly delay traffic flow and transports, causing increasing losses and slowing down the economic growth (Thurner 2001). Furthermore, time overrun means that the people (road users) and the economy have to wait for the road longer than it is necessary; which in turn limits the growth potential of the economy at large. On the other hand, the services provided by the road projects serve as input for many other sectors of the economy. Therefore, cost escalation leads to an increase in the capital-output-ratio for the entire economy. Combining these three major problems: quality shortfall, time overrun, and cost escalation, the result is reduction in the efficiency of the available resources and limited growth potential of the entire economy (Rwakarehe and Mfinanga 2013).

From the foregoing discourse, it can be noted that:

(a) Road projects have a close relationship to the national economy of the developed and developing countries.
(b) The increase in the paved roads is positively and significantly related to growth in the GDP.
(c) Road investments have a significant effect on the production sector's demand for labor, capital, and materials.
(d) Planning, designing, and construction of an integrated road network promotes sustainable development.
(e) Increasing the national's economy benefits of the road construction projects in Egypt on the can be achieved when these projects are delivered without quality shortfall, time overrun, and cost escalation.

3 Research Methodology

This paper presents the findings of a study carried out from October 2013 to October 2015 on the "successfully management of road construction projects in Egypt through identifying the factors influencing quality, time, and cost" for complementing the requirements for obtaining a Doctoral Degree in Civil Engineering, Faculty of

Engineering, Menoufia University, Shebin El-Kom, Egypt. The data was collected using questionnaires surveys, structured interviews, literature review, and case studies of road construction projects in the country.

4 Defining Quality Level, Time Overrun, and Cost Escalation

In order to recommend definitions for the terms of quality, time overrun, and cost escalation of construction projects specifically roads infrastructure, a comprehensive review of the previous studies regarding the iron triangle of projects management, management of road projects, quality, time overrun, and cost escalation of road projects have been conducted. This section presents the previous definitions of quality, time overrun, and cost escalation of the construction projects and the recommended definitions for these terms according to the authors viewpoints.

4.1 Quality Level

Deacon *et al.* (2001) defined the road construction quality as "*a major factor in determining how well a pavement will perform under traffic loading and when subjected to environmental influences*". On the other side, the Transportation Research Board Glossary included the following definition for quality of road as: "*(a) the degree of excellence of a product or service, (b) the degree to which a product or service satisfies the needs of a specific customer, and (c) the degree to which a product or service conforms with a given requirement*" (Committee on Management of Quality Assurance 2002).

Burati *et al.* (2003) presented three different definitions for quality as: "*(a) level of goodness, (b) customer satisfaction, and (c) conformance to requirements*". While, Mangold (2004) defined the quality of the pavement construction as "*quality of pavement equals quality of ingredients plus process quality. Quality of ingredient means quality of materials such as aggregate and asphalt. While process quality means quality of compaction, design mix, and equipments that have been used in the construction phase*".

Furthermore, Bhavanna Rao (2008) presented two definitions for the pavement quality. The first one was "*quality is generally understood to mean compliance with specified requirements*". While the second quality definition was "*quality is not merely conducting various tests on ingredients but quality is the totality of the characteristics of an entity that bear on its ability to satisfy the stated and implied needs of client/customer*".

As mentioned in the National Cooperative Highway Research Program (NCHRP) report No. 626 the pavement construction quality was defined as "*a tool or mean used by the owner and contractor to ensure that the desired results are obtained to produce high - quality and long- life pavements. Desired results are those that meet or exceed the specifications and design requirements*" (Transportation Research Board Executive Committee 2009). Moreover, Laughlin (2012) provided a definition for pavement construction quality as "*quality represents the activities and measures a contractor takes to manage a project and deliver the results that are specified in the contract and expected from the road users"*.

From the previous definitions for the pavement construction quality, it can be concluded that quality of road projects is "*the application of tests and approaches in the recommended specifications of the project to achieve the results of the state highway agency. These results are represented in getting adequate highway without any distresses until the end of the design lifetime of the road"*.

4.2 Time Overrun

Refer to the analysis of cost overruns and time delays of Indiana Department of Transportation (INDOT), Bordat *et al.* (2004) defined the time delay as "*the difference between a project's original contract period at the time of bidding and its overall actual contract period at the end of construction*". Also, Bordat *et al.* (2004) presented a formula in order to calculate the project time delay as follow:

$$\text{Time Delay} = \text{Last Day of Work} - \text{Expected Last Days of Work} \tag{1}$$

In the study of Al-Najjar (2008) the time overrun had been defined as "*the increased time to complete the project after planned date which caused by internal and external factors surrounded the project*". Also, Singh (2009) defined the time overrun as "*the difference between the actual and the initially planned (i.e., expected) dates of completion*". Tumi *et al.* (2009) defined the delay in construction industry as "*a critical function in construction projects*".

According to Ali *et al.* (2010) the definition of the delay in construction industry was "*a situation where the project can not be completed under the planned time. Also, it is a common issue faced in the construction industry all over the world especially in developing countries*". Also, construction delay was defined by Memon *et al.* (2011) as "*delays in progress compared to the baseline construction schedule*". In Baghdad, Jahanger (2013) defined the delay as "*the time over-run either beyond completion date specified in a contract or beyond the date that the parties agreed up on for delivery of project*".

Megha and Rajiv (2013) presented two definitions for the delay. The first one was "*the time overrun either beyond completion date specified in a contract or beyond the date that the parties agreed upon for delivery of a project*". While the second definition was "*a project slipping over its planned schedule and is considered as common problem in construction projects"*. In addition, Prakash Rao and Culas (2014) defined the delay of the constructed project as "*the late completion of works as compared to the planned schedule or the contract schedule*".

From the preceding discussion regarding the definition of delays, the time overrun of road projects can be defined as "*a conflict and the differences between the time in which the project is planned by the owner and the time in which the project is carried out by the contractor. This discrepancy or difference increases the execution duration of the project and means losing profits that are planned to be earned after starting the investment on the scheduled time which; in turn, limits the growth potential of the economy at large"*.

4.3 Cost Escalation

In the study of Bordat *et al.* (2004) a formula has been presented in order to calculate the project cost escalation as follow:

$$\text{Cost Escalation} = \text{Final Bid Amount} - \text{Original Bid Am} \tag{2}$$

Furthermore, Afetornu *et al.* (2006) defined the cost escalation of the constructed road projects in Ghana, as "*the additional cost beyond the planned estimated cost of the project*". While, Nega (2008) defined the cost escalation as "*the difference between the finial actual cost of a construction project at completion and the contract amount agreed by and between the client (the project owner) and the contractor during signing of the contract*". Matintupa (2009) known the cost escalation as an "*event occurs when the tenders for contracts clearly exceeded the client's own cost estimates of the construction costs*".

Refer to the analysis of cost deviations analysis in Palestinian road construction industry, Mahamid (2011) defined the cost escalation in road construction projects as "*a deviation that may be expressed as a percent difference between the final cost of the project (actual cost) and the contract award amount (estimated cost)*". Wakjira (2011) defined the cost escalation as "*the increase of the final actual cost of a construction project (usually expressed as a percentage of original contract amount) at a completion over the original contract amount, agreed by and between the client (the project owner) and the contractor during the signing of the contract*".

According to the theory of cost overrun in India, it was concluded that cost escalation is "*the ratio of the actual cost over the initially projected (i.e., expected) cost of the projects*" (Jain and Singh 2012). In Sri Lanka, Wijekoon and Attanayake (2012) known the cost escalation in road construction projects as "*a gap between the cost at completion and that originally estimated*". Moreover, Arcila (2012) defined the cost escalation in construction industry as "*a situation occurs when the final cost of the project exceeds the original contract value at the time of completion*".

From above, it can be said that cost escalation of road projects is "*a deviation in the target cost of the project. This deviation results from the lack of appreciation of the actual quantities that will be implemented in the project and the escalation of the materials prices, labors, and equipments during project execution. This deviation adds a significant financial risk to the economy; this in turn decreases the rate of national growth*".

5 Questionnaire Survey

The questionnaire survey, which has been designed to determine the percentages of quality level, time overrun, and cost escalation of road projects in Egypt, has been targeted at contracting companies specialized in the construction of road projects in Egypt. A systematic random sample has been selected to ensure a representative sample of all targeted respondents using Eq. (3) (Cochran 1977):

$$n = \frac{n_0}{1 + \frac{(n_0 - 1)}{N}} \tag{3}$$

where n, n_0, and N represent the sample size of the limited, unlimited, and available population, respectively. On the other hand, n_0 is estimated by the following equation:

$$n_0 = \frac{Z^2pq}{e^2} \qquad (4)$$

where Z is the statistical value for the confidence level, p is the value of the population proportion which is being estimated, q = 1 − p, and e is the desired level of precision. In the present research, a confidence level 90% is assumed thus Z = 1.65 from normality tables, p = 0.5 and e = (±10%). Substituting about: Z, p, q, and e in Eq. (4), the unlimited sample size of the population n_0 is approximated as follows:

$$n_0 = \frac{(1.65)^2 \times 0.5 \times (1 - 0.5)}{(0.10)^2} = 68.0625 \approx 69$$

Furthermore, the available population of the contracting companies of roads in Egypt were 2082 companies which were current members of the Egyptian Federation for Construction and Building Contractors (EFCBC) within all grades during conducting this research. The required representative limited sample size n of the construction companies has been determined using Eq. (3) as shown below:

$$n = \frac{69}{1 + \frac{(69-1)}{2082}} = 66.8 \approx 67$$

Based on the above equation, a total of 67 construction companies in Egypt were surveyed as a sample representing the available population of 2082 construction companies. The total number of completed questionnaires obtained from the 67 surveyed construction companies were 56 respondents.

5.1 Profile of the Questionnaire Respondents

This part discuss the designation and relevant experience of the respondents to the questionnaire. Figure 2 shows the designation of the respondents to the distributed questionnaire. The results of this Figure clarifies that 53.57% (30) of the contractors are site engineers, 17.86% (10) are project managers, and 28.57% (16) are office engineers.

Figure 3 shows the working experience of the respondents in the road construction industry, the analysis of this Figure clarifies that 39.28% of the respondents have experience in the field of road projects between 0 to 5 years, 25% have experience between 6 to 10 years, 16.07% have experience between 11 to 15 years, and 19.65% have experience more than 16 years.

The results of these Figures shows that the profile and experience of the respondents suggest sufficient exposure to make the information acquired reliable.

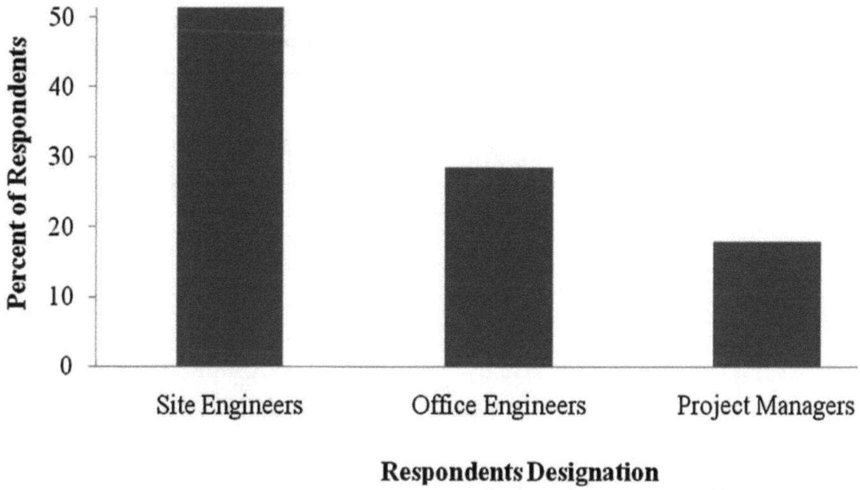

Fig. 2. Designation of the respondents

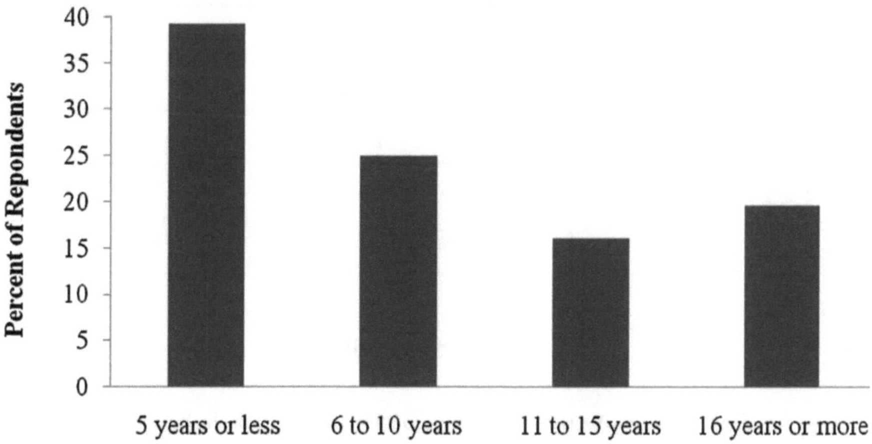

Fig. 3. Experience of the respondents in road construction industry

6 Analysis and Discussion of Questionnaire Survey Results

After collecting the raw data of the returned questionnaires, the succeeding stage is to analysis the survey results to obtain the percentages of quality level, time overrun, and cost escalation of road projects in Egypt. The results of the 56 questionnaires have been presented in Figs. 4, 5 and 6 which present the percentages of quality level, time overrun, and cost escalation respectively of 56 road projects in Egypt according to the experience of the participants.

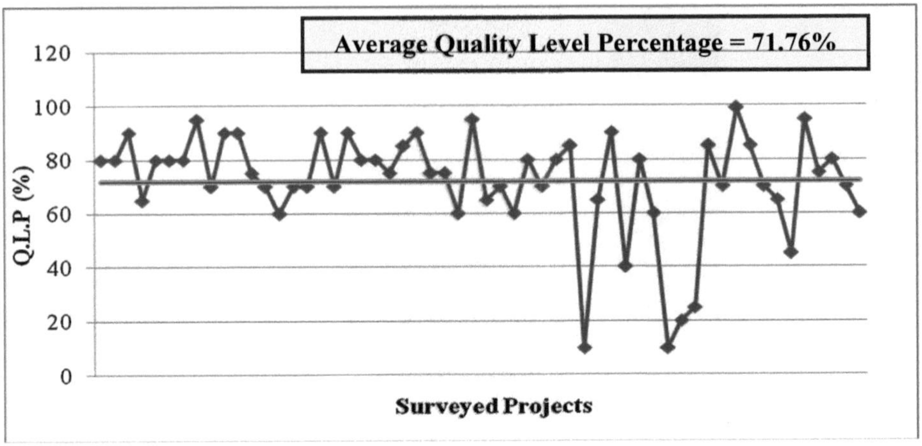

Fig. 4. Quality level percentage of 56 road projects in Egypt

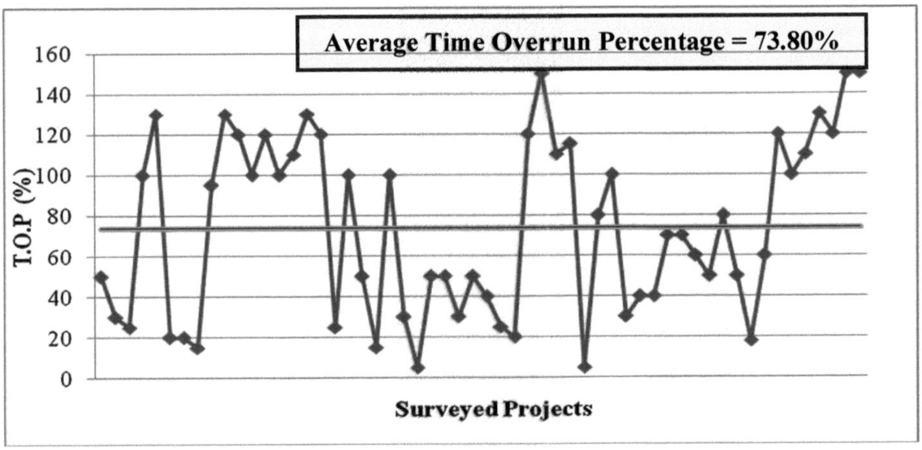

Fig. 5. Time overrun percentage of 56 road projects in Egypt

According to the shown Figures, the following findings have been concluded and summarized into the following points:

- The majority of road projects in Egypt, more than 60%, indicates that the quality level percentage ranges between 60% to 80%. Moreover, the average quality level percentage is 71.76% of the project target quality.
- The majority of road projects in Egypt, more than 75%, indicates that the time overrun percentage ranges between 20% to 120%. Moreover, the average time overrun percentage is 73.80% of the contractual project duration.
- The majority of road projects in Egypt, more than 70%, indicates that the cost escalation percentage ranges between 15% to 80%. Moreover, the average cost overrun percentage is 46.30% of the contractual project cost.

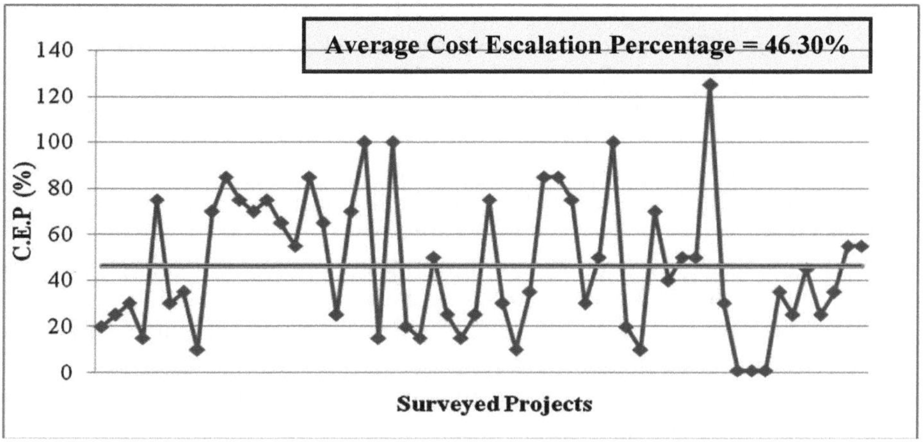

Fig. 6. Cost escalation percentage of 56 road projects in Egypt

7 Summary and Conclusions

Planning, designing, and construction of an integrated road network promotes sustainable development of the national economy. Furthermore, improving roads construction efficiency by means of quality effect, cost-effectiveness, and timeliness would certainly contribute to cost savings for the country as a whole. In an effort to improve the national economy of Egypt by improving its transportation system, this study has been conducted to find out more knowledge about the attributes of the iron triangle for projects management (quality, time, and cost), using road projects in Egypt as a case study.

According to the results presented in this study, the following conclusions can be drawn:

(a) According to the review of the previous definitions of the quality, time overrun, and cost escalation terms in many countries around the world, the authors concluded the following definitions for these terms in the roads construction industry as follows:

(a-1) Quality of road projects is: "*the application of tests and approaches in the recommended specifications of the project to achieve the results of the state road agency. These results are represented in getting adequate road without any distresses until the end of the design lifetime for the road*".

(a-2) Time overrun of road projects is: "*a conflict and the differences between the time in which the project is planned by the owner and the time in which the project is carried out by the contractor. This discrepancy or difference increases the execution duration of the project and means losing profits that are planned to be earned after starting the investment on the scheduled time which; in turn, limits the growth potential of the economy at large*".

(a-3) Cost escalation of road projects is: "*a deviation in the target cost of the project. This deviation results from the lack of appreciation of the actual quantities that will be implemented in the project and the escalation of the materials prices, labors, and equipments during project execution. This deviation adds a significant financial risk to the economy; this in turn decreases the rate of national growth*".

(b) According to the conducting field survey on the road projects in Egypt it has been noted that:

(b-1) The quality level percentage of the majority of highway projects in Egypt ranges between 60% to 80% with average percentage = 71.76%.

(b-2) All the surveyed projects suffering from time overrun issue with average time overrun percentage = 73.80%.

(b-3) All the surveyed projects suffering from cost escalation issue with average cost escalation percentage = 46.30%.

Refrences

Abedi, M., Fathi, M.S., Mohammad, M.F.: Effects of construction delays on construction project objectives. In: Proceedings of the 1st Iranian Students Scientific Conference in Malaysia, UPM, Malaysia, pp. 1–6 (2011)

Afetornu, C., Edum-Fotwe, F.T., McCaffer, R.: Estimating budget variability for road projects. In: Proceedings of the 22nd Annual ARCOM Conference, Birmingham, UK, pp. 563–571 (2006)

Akoa, B.B.: Cost overruns and time delays in highway and bridge projects in developing countries-experience from Cameron. M.S. thesis, Construction Management Program, School of Planning, Design and Construction, Michigan State University, East Lansing, Michigan, USA (2011)

Alhomidan, A.: Factors affecting cost overrun in road construction projects in Saudi Arabia. Int. J. Civil Environ. Eng. **113**(03), 1–4 (2013)

Ali, A., Smith, A., Pitt, M., Choon, C.: Contractors perception of factors contributing to project delay: case studies of commercial projects in Klang Valley, Malaysia. J. Des. Built Environ. **6** (02), 26–35 (2010)

Al-Najjar, J.M.: Factors influencing time and cost overruns on construction projects in the Gaza Strip. M.S. thesis, Faculty of Engineering, Islamic University, Gaza, Palestine (2008)

Arcila, S.M.: Avoiding cost overruns in construction projects in the United Kingdom. M.S. thesis, Warwick Manufacturing Group, University of Warwick, Coventry, Warwick, UK (2012)

Atkinson, R.: Project management: cost, time and quality, two best guesses and a phenomenon, its time to accept other success criteria. Int. J. Project Manag. **17**(06), 337–342 (1999)

Bhavanna Rao, D.V.: Quality control and assurance in road construction (2008). http://www.scribd.com/doc/6514982/Quality-Control-and-Assurance-in-Road-Construction. Accessed 24 Dec 2012

Bordat, C., McCullouch, B., Labi, S., China, K.: An analysis of cost overruns and time delays of INDOT projects. Publication No. FHWA/IN/JTRP-2004/07. Joint Transportation Research Program, Indiana Department of Transportation and Purdue University, West Lafayette, Indiana, USA (2004)

Burati, J., Weed, R., Hughes, C., Hill, H.: Optimal procedures for quality assurance specifications. Report No. FHWA-RD-02-095, Federal Highway Administration, Washington, D.C., USA (2003)

Chitkara, K.: Construction Project Management, Planning, Scheduling and Controlling, 4th edn. Tata McGraw Hill, India (2004)

Cochran, W.G.: Sampling Techniques, 3rd edn. Wiley, New York (1977)

Committee on Management of Quality Assurance: Glossary of highway quality assurance terms. Transportation Research Circular No. E-C037, Transportation Research Board, Washington, D.C., USA (2002)

Deacon, J.A., Monismith, C., Harvey, J., Popescu, L.: Pay factors for asphalt-concrete construction: effect of construction quality on agency costs. Technical Memorandum, TM-UCB-PRC-2001-1, Pavement Research Center, Institute of Transportation Studies, University of California, Berkeley, California, USA (2001)

Enshassi, A., Al-Hallaq, K., Mohamed, S.: Causes of contractor's business failure in developing countries: the case of Palestine. J. Constr. Dev. Ctries. **11**(02), 1–14 (2006)

Falqi, I.I.: Delays in project completion: a comparative study of construction delay factors in Saudi Arabia and the United Kingdom. M.S. thesis, School of the Built Environment, Heriot-Watt University, UK (2004)

Fan, S., Kang, C.C.: Road development, economic growth, and poverty reduction in China. Research Report No. 138, International Food Policy Research Institute, Washington, D.C., USA (2005)

Farag, I.: SWOT Analysis of Road Transport Sector in the Arab Region. IRU Academy, Alexandria (2012)

Giridhar, P., Ramesh, K.: Effective management of turnkey projects. In: AACE International Transactions, pp. PM.04.1–PM.04.5 (1998)

Hendrickson, C., Au, T.: Project Management for Construction: Fundamental Concepts for Owners, Engineers, Architects and Builders, 1st edn. Prentice Hall, Upper Saddle River (1989)

Ibn-Homaid, N.T., Eldosouky, A.I., Al-Ghamdi, M.A.: Change orders in Saudi linear construction projects. Emir. J. Eng. Res. **16**(01), 33–42 (2011)

Jahanger, Q.I.: Important causes of delay in construction projects in Baghdad city. Aust. J. Basic Appl. Sci. **07**(04), 14–23 (2013)

Jain, N., Singh, R.: A theory of cost overruns. In: Proceedings of the 8th Conference of Economic Growth and Development, Indian Statistical Institute, New Delhi, India (2012)

Kaliba, C.: Cost escalation, schedule overruns and quality shortfalls on construction projects. M. S. thesis, Department of Civil and Environmental Engineering, School of Engineering, University of Zambia, Lusaka, Zambia (2010)

Kaliba, C., Muya, M., Mumba, K.: Cost escalation and schedule delays in road construction projects in Zambia. Int. J. Proj. Manag. **27**(05), 522–531 (2009)

Laughlin, R.: Questioning quality. http://www.retailfacilitybusiness.com/industry-voices/534-questioning-quality-industry-voices.html. Accessed 13 Oct 2013

Mahamid, I.: Effects of project's physical characteristics on cost deviation in road construction. J. King Saud Univ. – Eng. Sci. **25**(01), 81–88 (2013)

Mahamid, I.: Analysis of cost deviations in road construction activities: a case study from Palestine. Jordan J. Civil Eng. **05**(04), 552–567 (2011)

Mangold, P.E.: Batching, mixing, transport and placement methods for quality concrete pavements. Presentation at Transportation 2004 Workshop, Fort Lauderdale, Florida, USA (2004)

Matintupa, E.: Impact of the planning processes on the road construction costs. In: Proceedings of the 27th International Baltic Road Conference, Riga, Latvia, pp. 1–8 (2009)

Megha, D., Rajiv, R.: A methodology for ranking of causes of delay for residential construction projects in Indian context. Int. J. Emerg. Technol. Adv. Eng. **3**(03), 396–404 (2013)

Memon, A., Abdul Rahman, I., Abdullah, M., Azis, A.: Assessing the effects of construction delays on MARA large projects. In: Proceedings of International Conference on Advanced Sciences, Engineering and Information Technology, Malaysia, pp. 624–629 (2011)

Nadiri, M.I., Mamuneas, T.P.: Contribution of highway capital to output and productivity growth in the U.S. economy and industries. Department of Transportation, Federal Highway Administration, USA (1998)

Nega, F.: Causes and effects of cost overrun on public building construction projects in Ethiopia. M.S. thesis, Faculty of Technology, Addis Ababa University, Addis Ababa, Ethiopia (2008)

Oberlender, G.: Project Management for Engineering and Construction, 2nd edn. McGraw-Hill, New York City (1993)

Ozdemir, M.: A probabilistic scheduled delay analysis in construction projects by using Fuzzy logic incorporated with relative important index (RII) method. M.S. thesis, Civil Engineering Department, School of Natural and Applied Sciences, Middle East Technical University, Turkish (2010)

Prakash Rao, B., Culas, J.C.: Review article: causes of delays in Construction projects- a case study. Int. J. Current Res. **6**(06), 7219–7222 (2014)

PWC's Staff Reporter: Africa gearing up-future prospects in Africa for the transportation & logistics industry (2013). http://www.pwc.com/africagearingup. Accessed 13 Mar 2015

Rwakarehe, E., Mfinanga, D.: Effect of inadequate design on cost and time overrun of road construction projects in Tanzania. KICEM J. Constr. Eng. Proj. Manag. **04**(01), 15–28 (2013)

Singh, R.: Cost and time overruns in infrastructure projects: extent, causes and remedies. Working Paper No. 181, Centre for Development Economics, Department of Economics, Delhi School of Economics, University of Delhi, Delhi, USA (2009)

Thurner, H.F.: Quality assurance and self control in road construction, advanced measurement technology. Technical Paper, Geodynamic AB, Stockholm, Sweden, pp. 1(11)–11(11) (2001)

Timetric: Construction in Egypt - key trends and opportunities to 2016 (2012). http://www.marketresearch.com/Timetric-v3917/Construction-Egypt-Key-Trends-Opportunites-7084362. Accessed 23 Oct 2013

Transportation Research Board Executive Committee: NDT technology for quality assurance of HMA pavement construction. NCHRP Report No. 626, Transportation Research Board, Washington, D.C., USA (2009)

Tumi, S.A., Omran, A., Pakir, A.H.: Causes of delay in construction industry in Libya. In: Proceedings of the International Conference on Economics and Administration, Faculty of Administration and Business, University of Bucharest, Romania, pp. 897–906 (2009)

Wakjira, T.: Risk factors leading to cost overrun in Ethiopian federal road construction projects and its consequences. M.S. thesis, Department of Civil Engineering, Addis Ababa Institute of Technology, Addis Ababa University, Addis Ababa, Ethiopia (2011)

Wijekoon, S.B., Attanayake, A.M.: Study on the cost overruns in road construction projects in Sri Lanka. In: Proceedings of the 2nd International Conference on Sustainable Built Environment, Kandy, Sri Lanka, pp. 111–121 (2012)

Amaravati - A City Reborn, Journey Towards a World-Class Smart City

Madhusmita Ghadei(✉)

National Institute of Technology, Karnataka, Mangalore, India
madhusmita.ghadei26@gmail.com

Abstract. Amaravati, a small village located in Guntur district, remembered vaguely only for its history became the talk of the nation as soon as it was sanctioned to become the next capital of Andhra Pradesh, along with being chosen in the 2nd phase for the ambitious Smart Cities Mission undertaken by the Govt. of India. India is a fast developing economy and projects such as these are furthering its growth. The plan was presented by the Govt. of Singapore to the Andhra Pradesh Chief Minister, N Chandrababu Naidu. It is a 5-phase plan sprawled over 16.9 sq.kms. which mainly concentrates on "providing jobs and homes, world-class infrastructure, quality living, protecting identity and heritage and resource management and environment." The SEED Capital Area (SCA) Master Plan aims to implement a well-connected network of transport hierarchy, effective disaster management, innovative waste management systems paving way for a greener tomorrow and sustainable, futuristic infrastructure based on sustainable development strategies. This paper covers the planning (SEED Master Plan), current progress and projected goals along with the importance of such a formidable plan for the state of Andhra Pradesh and a country like India.

1 Introduction

India, an integral part of BRICS (acronym for Brazil, Russia, India, China and South Africa which are deemed as the association of five major emerging national economies), the 7th largest economy of the world, with a phenomenal GDP growth rate of 7.5% as of 2015 has been consistently developing towards becoming a full-fledged super power. The democratic government has contributed greatly towards this growth. One such initiative was undertaken by the Government of India with key people such as Venkaiah Naidu and Narendra Modi, in 2015. A total of 100 cities from the various states in India had to undergo a rigorous multi-stage selection process in order to get funding. Amaravati although not technically a city, entered this competition and cleared the 2nd phase selection process and hopes are high that it gets sanctioned to be chosen and funded under the Smart Cities Mission.

Andhra Pradesh, one of the major states in the southern part of India underwent bifurcation in June 2014 giving rise to the new state of Telangana. While the previous capital of Andhra Pradesh, that is, Hyderabad continued to serve as capital for Telangana, Andhra Pradesh was left without one, which led to the need for a new capital and following certain disagreements, none of the existing cities fit the profile, hence a new

J. Calautit et al. (eds.), *Towards Sustainable Cities in Asia and the Middle East*,
Sustainable Civil Infrastructures, DOI 10.1007/978-3-319-61645-2_2

city was sanctioned to be designed and constructed. The area selected for this project was located between the two biggest regions of Andhra Pradesh, that is, Guntur and Vijayawada. Thus the dream of Amaravati was born. Amaravati, believed to be the regional center during the Mauryan rule and later the capital of the Satavahanas dynasty soon lost its stature with modernization and remained relevant only in terms of historical importance. The selected region consists of a large number of villages and is one of the biggest and most successful examples of Land Pool Scheme wherein the inhabitants of the collective area gave up their small pieces of land to the government in return for compensation, thus enabling the formation of a uniform and well-designed skeleton.

The project has consistently been in the media limelight due to continued efforts by Chief Minister of the state, N Chandrababu Naidu who came up with various novel techniques to achieve just that, one being a very elaborate event dedicated to laying foundation, wherein holy water from different pilgrimage sites all over the world was poured on the foundation stone and the event was attended by important government officials and dignitaries from India. Another interesting scheme used was "My Bricks – My Amaravati" through which the general public could contribute towards building the state capital by buying these virtual "bricks" each costing a meagre INR Rs. 10 thereby further funding the project of making Amaravati a world class city.

An MoU was signed between the Singapore and Andhra Pradesh government on December 8, 2014 to collaborate on conceptualizing the design for the capital city. Surbana International Consultants Pte Ltd. and Jurong Consultants Pte Ltd. were hired to do the same, while L&T, one of the largest construction companies in India was assigned to implement it. In January of 2016, the Andhra Pradesh Capital Region Development Authority (AP CRDA) kicked off the Amaravati Government Complex Concept Design Competition which saw participation from three Master Architects to come up with a lean and unique design for the capital, and was concluded in March, 2016 with Fumihiko Maki of Maki & Associates winning the competition.

This ambitious project has had its shares of ups and downs, with some calling it "too good to be true" but is built on the dreams of millions who gave up a lot to contribute towards making this a reality, while the challenges are many, the goals are not unattainable, the detailed planning, execution timeline, achieved targets, etc. will be discussed in the following sections.

2 Site Selection

Andhra Pradesh is the 8th largest state in India covering an area of over 160,000 km and surrounded by the Indian states of Karnataka, Telangana, Odisha, Chhattisgarh and Tamil Nadu with Bay of Bengal terminating the land area in the east. The major cities of Andhra Pradesh (AP) are Vishakhapatnam, Vijayawada and Vizianagaram to name a few. The reason for not choosing any of the pre-existing cities was two-fold. One was due to the conflict amongst the masses each claiming that their city deserved to be the new capital and other was due to accessibility, the APCRDA and the Hon'ble Chief Minister desired the capital to be centrally located which gave rise to the need for a new city and a new capital as a whole.

Table 1. Existing land distribution

Land use	Area (ha)	%
Developable land	1449.29	85.55%
Village settlements	29.40	1.74%
Island	182.13	10.75%
River Krishna	31.80	1.88%
Water bodies	1.43	0.08%
Total	1694.05	100%

The site selected was that of Amaravati, flanged on either side by the two major regions of Vijayawada in the north-west and Guntur in the south. It was nothing but a collection of small villages, and rural landscapes mainly uninhabited (Refer Table 1 for the Existing Land Usage). Post the hugely successful land pooling scheme, the area acquired was over 200 kms and is a riverfront site facing Krishna river in the north.

Within the proposed city, the large north-south axis was chosen, to provide grandeur and ceremonial quality to the government house and also to be closer and easily accessible to the neighbouring city of Vijayawada with a large population of over a million while retaining its own identity as the new state capital. Neerukonda hills will terminate this axis on the southern end while the northern axis appears to extend endlessly to the Kandapalli Hill forest. (Refer Figs. 5 and 6 for available site options and final selected site).

In January 2015, Surbana and Jurong teams undertook a 2 days elaborate site visit to get a first-hand feel of the site and surroundings. The team also interacted with local people to understand their concerns and issues to be considered during the formation of the master plan. All this information was utilized in the site analysis and formation of the plans (Table 2).

Site Selection Criteria:

- Proximity to Krishna river, one of the major Indian rivers which gives the capital its stature of riverfront capital.
- Flood free zone, since the land selected is mainly flatlands hence is least prone to flooding and does not necessitate drastic measures to combat the same.
- Proximity to Vijayawada, the 2nd largest city in the state in terms of GDP and a population of over a million. Ease of access is increased but at the same time the individual identity of the state capital is retained.
- Blank canvas, since the area selected has a rather small population of just 1.74% of the total land area, the site could be developed appropriately without worrying about pre-existing systems and without affecting livelihood, businesses and culture of the inhabitants.
- Vaastu, an important aspect of construction and architecture in India which was a part of ancient texts in India detailing the spatial/geometric layout of designed site along with placement of various integral structures in a house, building or region. The north-south axis is considered most auspicious and hence will form the core of the government in accordance with Vaastu Shashtra principles.

Table 2. SEED: landuse distribution table

Sl. no.	Landuse	Total area (ha)	Percentage (%)
1	Business park	39	2.30%
2	Commercial	47	2.80%
3	Existing habitation	29	1.74%
4	Government administration	50	3.53%
5	High density residential	45	2.63%
6	Infrastructure	11	0.64%
7	Institution	17	0.99%
8	Low density residential	84	4.99%
9	Medium density residential	52	3.69%
10	Mixed use developments	255	15.07%
11	Neighbourhood centre	4	0.23%
12	Parks and greens	331	19.96%
13	Roads	277	16.37%
14	Schools	14	0.82%
15	Special use	119	7.01%
16	Special development zone	225	13.27%
17	Village buffer	4	0.22%
18	Water	70	4.13%
	Total area	1694	100%

- Scenic beauty, the location is one of the most aesthetically pleasing ones with a good number of natural formations, historical structures and vantage points with great view of the landscape and would serve as a good tourist attraction garnering significant tourist revenue.
- Good connectivity, since it is located between Vijayawada and Guntur district.

Although the challenges posed were minimal, the main ones were the hindrance in optimal use of the inhabited area of 1.74% which is but a minor problem. The bund located at a distance of 250 m from the water edge needs to be realigned in order to continue the development without any obstacles. Apart from these, the site selection has proven to be very successful and hassle-free thereby ensuring speedy completion of the project.

3 Methodology

Surbana and Jurong teams collected the necessary data from the Andhra Pradesh agencies during the period of December 2014 and May 2015. Using this data, the teams filtered out specific relevant data that can be applied in the development of the Capital City and SEED development Master Plans. The data was received in several formats including GIS, Autocad, Excel sheets and hard copy reports (Fig. 1).

Several meetings were held during the team's stay in Hyderabad that included CRDA officials and Dr. P. Narayana, Minister of Municipal Administration & Urban Development, Urban Water Supply and Urban Planning for Andhra Pradesh. The team

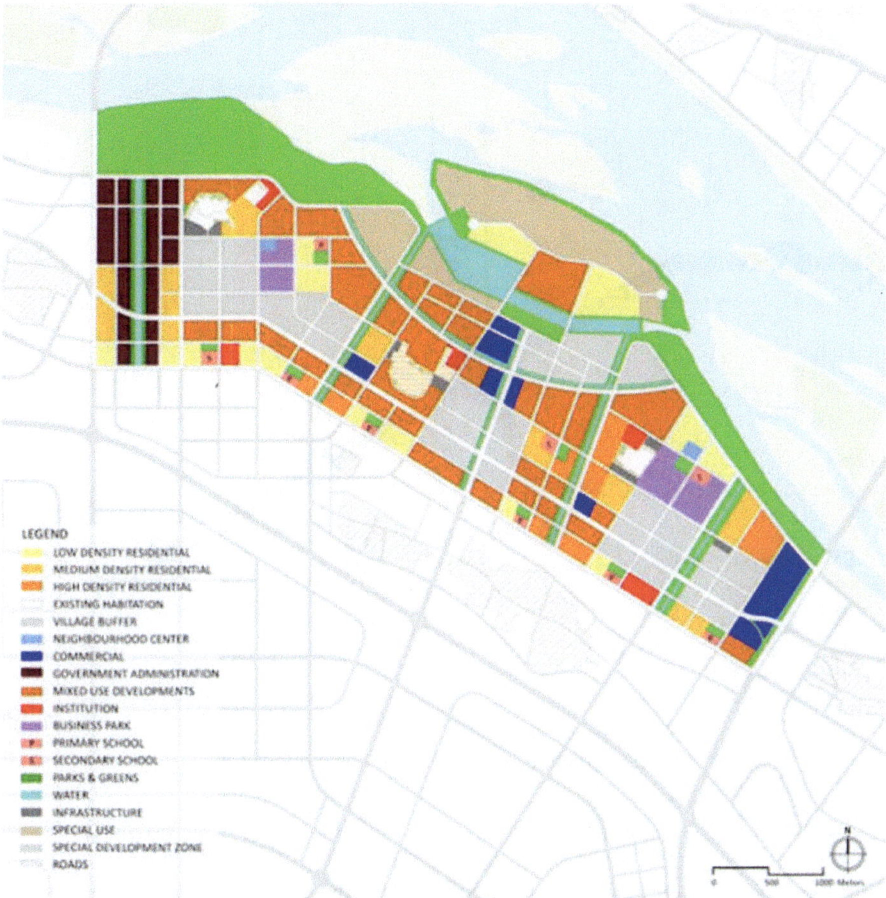

Fig. 1. SEED MasterPlan

from Singapore included officials from Surbana International Consultants Pte Ltd. (Surbana), JURONG Consultants Pte Ltd. (Jurong), Centre for Livable Cities (CLC), Ministry of Trade and Industry (MTI), Singapore Cooperation Enterprise (SCE) and International Enterprise (IE).

A delegation of 23 Andhra Pradesh officials attended the APLUGP (Andhra Pradesh Leaders in Urban Governance Programme) organized by CLC in Singapore from 19th to 24th January. Surbana and Jurong teams were actively involved throughout this workshop. Surbana and Jurong presented and discussed the progress of the Master Plan with the delegation members to get their insights and views on the ideas being considered.

Parallel to this, planners from Surbana and Jurong were analyzing the relevant data received from the Andhra Pradesh agencies. The analysis included understanding of physical site features such as hills, rivers, canals, forests, drains, etc., along with an understanding of the socio-demographic and economic profile of the Capital City and the SEED development area.

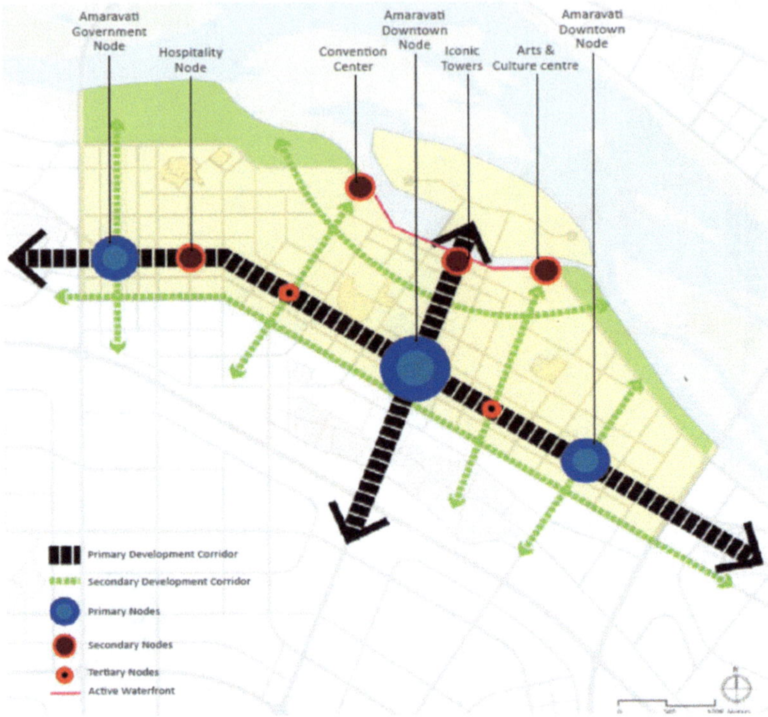

Fig. 2. Structure plan for SEED development area

4 Smart City

While there is no fixed definition for a smart city, the most vital step for a city to be considered smart is its infrastructure which needs to be prepared to handle the population explosion efficiently, be disaster proof, constructed keeping in mind the pressure on public transport, and in general be superior in every aspect as compared to the present city designs.

Some well-known smart cities in the world are Stockholm, Southampton, and Amsterdam. India too is attempting to jump on this wagon by implementing the Smart Cities Initiative wherein 100 cities selected from all over India would be funded to be converted into a smart city. The procedure consists of rigorous inspection and competition at various levels to qualify for the funding. Amaravati, technically not an existing city of India was not selected for this prestigious status in the first two rounds but hopes are high that it will meet the standards soon. Despite that, the program is primarily funded by the central government aside from state funding and foreign investments.

The city is designed to be built on a Smart City Framework (Refer Fig. 3 at the end of the paper) with six primary goals, each corresponding to the three pillars of sustainability. Establishing this framework is the most critical step preceding the Vision as this guides the Goals and Strategies. It lays out the Key Performance Indicators to address the social, economic and environmental issues specific to the focus area.

Fig. 3. Smart city framework

1. ECONOMIC
   - Jobs & Homes for All
   - World Class Infrastructure
2. SOCIAL
   - Identity & Heritage
   - Quality Living
3. ENVIRONMENT
   - Clean and Green
   - Effective Resource Management

To achieve the above mentioned goals, the planning has been done as follows:

1. 4 hierarchies of commercial and 5 residential developments have been proposed.
2. A well-woven network of parks and open spaces running across the capital and terminating at the riverfront.
3. A well connected road and public transport network is in works.
4. The iconic sites of the city have been restored and well-maintained to preserve the heritage of the capital.
5. Waste water management, flood prevention, solid waste management, water and power supply methodologies to be implemented throughout Amaravati.

4.1 Residential Plans

5 residential developments are offered to provide a variety of lifestyle to the public.

1. High Density Residence: This is expected to house almost 15% of the total population and is proposed along the river Krishna waterfront and along the linear parks lining the waterfront. The type of building suggested is high rise apartments (of up to 20 stories) which are an effective way to deal with the land availability problem in India and are quite popular among middle class families. The apartments would also be integrated with community centers, playground, club house, pools, etc. giving the feeling of a typical gated community.
2. Medium Density Residence: This would house a smaller percentage of the population and is expected to be more affordable as, with common facilities within the apartment compound and built along green corridors. The height would be restricted to 8 stories and about 30% of the apartments would be meant for government employees and their families.
3. Low Density Residence: These would consist of detached and semi-detached houses limited to two-stories. Although land occupancy versus families housed ratio would be high, this residential plan would appeal to public that can afford individual housing and would prefer decongested residences. A set of luxury villas are also proposed on the island to limit the population residing there.
4. Mixed Use Residence: This type of residential space is expected to accommodate almost 60% of the SEED population and is built on the principles of "live, work and play" with a 70%–30% combination of residential and commercial spaces respectively. It would be built along the transit nodes to maintain high mobility in the region, even post office hours. A height of about 25 stories is set as upper limit for these apartment complexes.
5. Village Residence: Amaravati before being transformed into a world-class smart city consisted of small village tracts and that is exactly what this type of residency is trying to preserve, staying true to the character of this ancient city. Three villages would be retained as they are, each with unique characteristics situated near the Government Core, Downtown and near the arts and cultural hub of the capital.

4.2 Transportation

The road network is very well-planned for the capital in order to meet the demands of the expected and projected population with a hierarchy of transport layers each with a unique role (Refer Fig. 4 for SEED Road Network Plan). Out of the 88 km of road planned, 1.5% is for expressway, 5.7% for arterial roads, 8% downtown, 24% sub-arterial and 60% collector roads.

1. Urban Expressways: Built on the eastern edge of the development, this is expected to serve as the entrance to the city, consisting of the iconic gateway bridge. Also, it would serve as a connected to the neighbouring city of Vijayawada and Gannavaram airport.

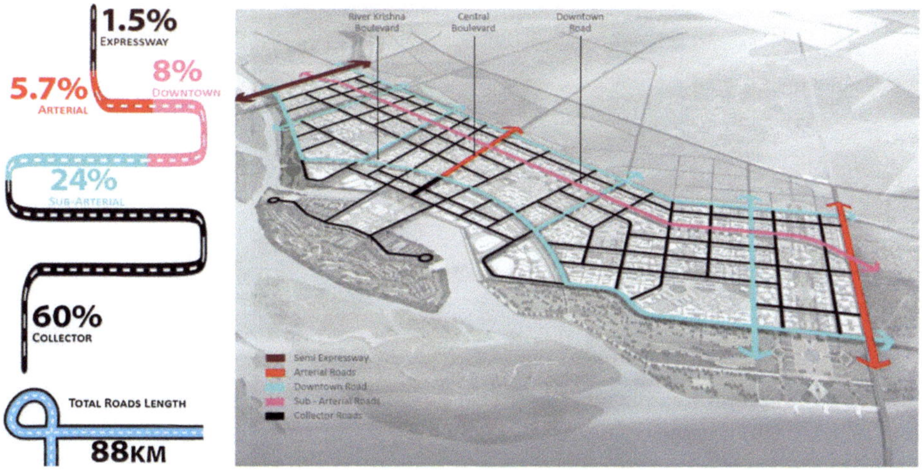

Fig. 4. SEED road network plan

2. Major Arterial Road Network: Responsible for carrying major vehicular traffic and mass transit, this arterial network would connect the government core to NH9, and residential areas. Also, it would connect to the newly proposed airport suggested to be built on the south region of the capital.
3. Sub-Arterial Road Network: Sub-arterial roads by definition, support the arterial roads and are responsible for providing access to chunks of the city, built along the green transit.

The public transport network is meant to encourage the public to travel by public transportation rather than resort to private vehicles, thereby reducing traffic and stress on the roads. It is built on three main network idea, i.e., Mass Rapid Transit Networks (MRT), Transit Oriented Development (TOD) and Bus Rapid Transit Loop.

The typical Character Roads defined by the type of building it is surrounded by and the public are: Downtown Road, Central Boulevard, River Krishna Boulevard and Collector Roads.

4.3 Amenities

4.3.1 Flood Management

Since the capital is planned along a huge water body, the risk of flooding exists and could be due to overflowing of Krishna River or runoff from Kondaveeti Vagu which is a tributary of Krishna River. Based on previous data and reports, the maximum rise of water is expected to be around 20 m above Mean Sea Level (MSL) while the bunds constructed to protect the capital range from 25 to 27 m above MSL.

To combat the flooding situation, the solution provided is simple yet elegant:

- Realigning and strengthening the river by lowering the existing bund platforms.
- Raising of platform on the island to prevent the water from entering the infrastructure in case of flooding.
- Construction of diversion channels to direct the water into Krishna river.

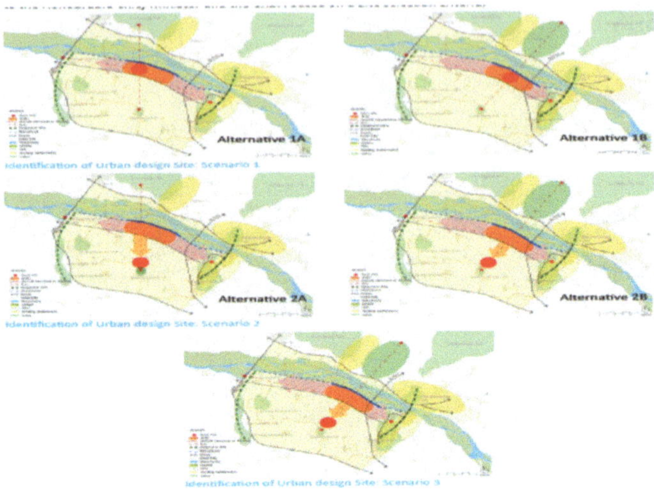

Fig. 5. Site options available

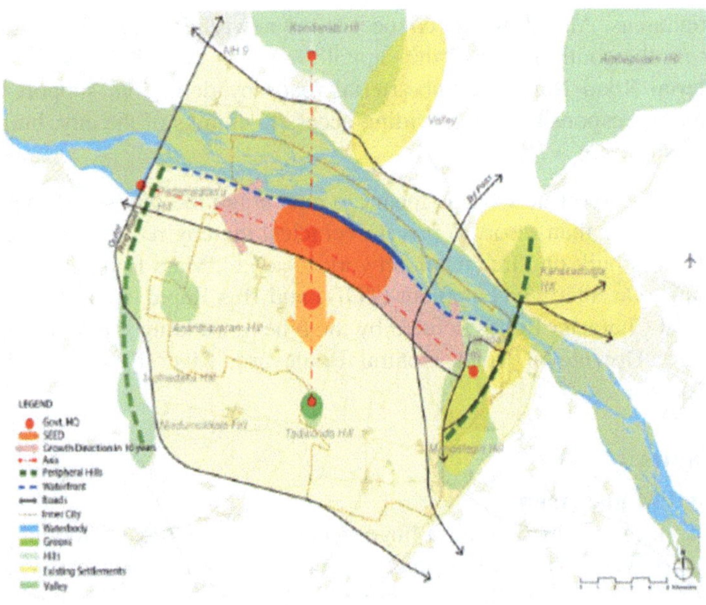

Fig. 6. Selected site

4.3.2 Water

Since a proper water supply plan and infrastructure is not already in place at the capital, a new one has to be built from scratch to meet the water needs of the city. (Refer Table 3 for Water Demand Projections).

Table 3. Water demand projections

	Water demand (MLD)	Service zone
Phase 1	11.6	WDC1
Phase 2	26.1	WDC1
Phase 3	26.1	WDC2
Phase 4	13.6	WDC3
Phase 5	18.3	WDC3
Total	95.7	

Water Treatment Plant would be built in stages but needs to be complete before Phase 1 of the SEED plan is implemented, it can be expanded based on the demand. It will draw water from the Krishna River.

Water Transmission Mains would be the means of transporting the treated water and would be built along the Downtown Road. The transmission mains need to be laid down at the same time as the road network so as not to disturb the infrastructure in case expansion is required.

Water Distribution Centers would collect water from the Water Main and store for later supply.

The waste water can be dealt with the temporary sewage treatment plant (STP) or gravity trunk sewer. (Refer Table 4 for Sewage Treatment Projections) Temporary sewage water treatment plant would be constructed outside the SEED development and its capacity would increase with the implementation of successive phases. Gravity Trunk Sewer would run along the downtown road connecting to the temporary STPs.

Table 4. Sewage treatment projections

	Sewage projections (MLD)
Phase 1	9.3
Phase 2	20.9
Phase 3	20.9
Phase 4	10.9
Phase 5	14.6
Total	76.6

4.3.3 Solid Waste Management

As a temporary solution, the landfill site at the neighbouring city of Guntur could be utilized to dispose of the solid waste generated. Instead of relying on heavy infrastructure, it is planned to construct a simple integrated solid waste management system in Vijayawada, wherein waste could be converted to energy.

5 Green Amaravati

Pollution has been a major problem in India for decades now, and keeping that in mind, Amaravati tries to curb it before it becomes serious. One simple way to achieve this is by having protected green belts, forested regions and multiple green parks spread throughout the city. The civic axis is meant to be a ceremonial one running along the downtown road all the way to the government core. It gives a sense of easy accessibility to the core while at the same time serving as a gathering ground for the government and public, useful in times of parade as well. (Refer Fig. 7).

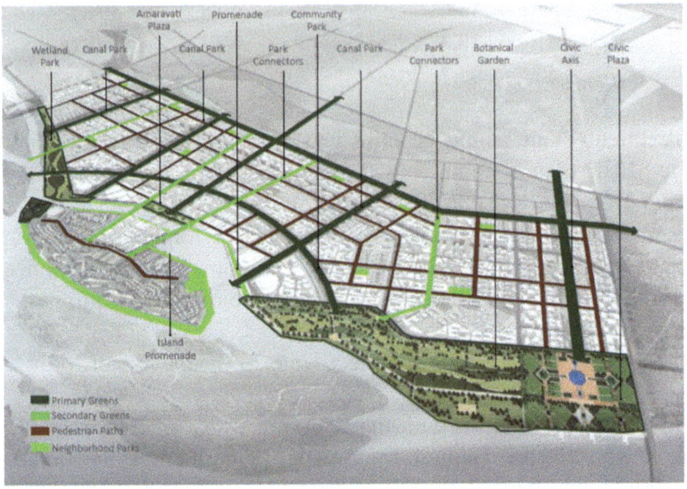

Fig. 7. Green Amaravati

The canal parks would be an extension of the arterial roads, meant to promote recreational activities within and along the canal. They serve a dual purpose because in case of flooding, the can act as natural drains into the Krishna River.

Community parks would line the river Krishna having a unique arc-shape complimenting the concave of the river.

The Botanical Garden is planned to skillfully display the existing flora and fauna of the region, and would be a central attraction for the public.

The Wetland Park would seamlessly merge the beauty of the waterfront with that of the vegetation. It would serve as a picnic, camping, jogging spot and would be a characteristic feature of the green Amaravati plan.

All of the above mentioned plans would serve as primary greens. Secondary greens will act as park connecters and shaded walkways and promenades which would beautifully tie together the waterfront greens.

The tertiary network would connect the primary greens to nearby parks. All in all, a total of 282 HA of land would be dedicated to primary greens, 69 HA to secondary greens and neighbourhood parks and 20.6 km to green pedestrian walkways.

6 Place-Making Strategies

4 key nodes of development have been planned, each dealing with a different aspect of the city life (Refer Fig. 2 for Structure Plan for SEED Development Area):

1. Government Core: This is the whole reason for construction of the capital. It would consist of the three important arms of the government viz. The Secretariat, The Assembly and The High Court. Along with all the official space, accommodation for the office bearers and their families would also be provided. A large green central region would serve as a center for F&B activities, the aim being to make the core as accessible as possible to the public but not at the cost of security. Vehicular traffic will need to be diverted and must be minimized around the Core.
2. Amaravati Gateway: It would be the eastern entrance to the city consisting of the iconic gateway tower and iconic bridge providing the incoming tourists with a sense of grandeur as soon as they approach the city. The gateway would also be rife with a lot of open green places further emphasizing the beauty of the entrance.
3. Amaravati Downtown: As per the SEED Master Plan, the downtown is the commercial heart of the capital which would see maximum activity in terms of community involvement. It is planned to be well-connected by all public transit with the interchange of two transit lines at the central TOD serving as the high density commercial hub consisting of parks, canals, central boulevard, downtown road, etc.
4. Amaravati Waterfront: The waterfront is planned to be utilized to its full extent in terms of commercial and recreational usage. Community centers, parks, walkways, amphitheaters, etc. would all serve to attract the public. At the same time, the waterfront would be advertised as a high-end residential plot to maintain the serenity of the region by reducing congestion.

7 Implementation

The execution of this master plan is set to be in 5 major phases. (Refer Table 5 for Phase-wise Development Plan)

Table 5. Phase-wise development plan

	Total seed area	Residential population	Gross floor area (mil. m^2)	Total jobs
Phase 1	18%	26,000	2.49	95,000
Phase 2	18%	94,000	4.03	160,000
Phase 3	18%	99,000	4.04	152,000
Phase 4	11%	29,000	3.29	177,000
Phase 5	35%	50,000	3.35	113,000

Phase 1: The ultimate goal of building this new city is so that the newly formed government can immediately shift its headquarters to Amaravati and that is what the Phase 1 aims to implement. Hence development would start from western end of the region, with transport and downtown road network given high priority. The pre-existing village settlements would be retained as is.

Phase 2: Once the government core is complete, the growth would naturally be eastward with more emphasis on residential settlement. With growth in residents, pressure on road networks would also be high and hence the Bus Rapid Transit system (BRT) has to set in place. Other key developments include botanic Gardens on the river Krishna bank.

Phase 3: Following Phase 2 development, Phase 3 would grow in a similar fashion as that of Phase 1. BRT would be expanded and the arc road and collector road would be developed and completed in this phase. This phase would also concentrate on development of the commercial center and eastern business park.

Phase 4: The iconic gateway bridge and towers, in essence, the eastern entrance to the capital would be completed in this phase. The pressure on traffic is expected to rely completely on Mass Rapid Transport (MRT) corridors by the end of this phase. The city wetland park, complementing the botanical gardens would also be completed in this phase.

Phase 5: This phase would concentrate more on the development along the waterfront, that is along Krishna River, completing the island cluster. Since most of the roads and transit networks would be completed in the previous phases, only intimate road networks would have to be built based on the growing demands, if any.

8 Conclusion

Amaravati was envisioned to be a dream city for the state of Andhra Pradesh, it is one of the most ambitious infrastructure development projects taken up by the government. As per planning Phase 1 is expected to be completed by 2018, and progress is being made in that direction. The new Andhra Pradesh government will be moving to the partially complete core on the auspicious day of Dushera, an important festival in India. A lot of talk has been rife regarding the plausibility of completing a project of this magnanimous scale within the estimated timeline, and many have written off the realization of such tall claims, but the work being done has managed to keep the critics at bay at least for the time being. The capital Amaravati is constantly on the news, scrutinized by the media and public alike with respect to the progress, which perhaps serves as another incentive for the parties involved to ensure timely completion of the development.

Although the government of India launched the Smart Cities Initiative in 2015, Amaravati is not yet officially a part of the selected cities, hence the funding comes from different international investors, to whom the government of Andhra Pradesh is answerable. Amaravati is already pioneering the new wave of emerging smart cities in India and hopefully would do a good job of the same.

Acknowledgments. First of all, I would like to thank God Almighty for giving me this unique and wonderful opportunity, it is way more than anything I could have anticipated.

I wish to express my sincerest gratitude to the professors especially Prof. Dr. Varghese George and HoD Prof. Dr. D.V. Reddy from Department of Civil Engineering, NITK, Surathkal for inspiring me and motivating me to take up new challenges every day and give it my best shot.

I owe a huge gratitude to the SUCI Editorial team and GeoMEast panel for short-listing my abstract, and making me believe that nothing is impossible. I never would have dreamed of landing something so big and I couldn't be more thankful. The constructive criticisms and feedback provided by the panel of reviewers also helped a great deal.

Last but, definitely not the least, I would like to pay my gratitude to my beloved parents for their trust, help and incessant support throughout this endeavour. Their continual faith in me is what kept me going. I would also like to thank my friends for their continuous support and encouragement and also for greatly contributing towards proof-reading the paper.

Improving Sustainable Construction Practices Through Facility Management

Clinton O. Aigbavboa(✉), Ayodeji E. Oke, and Looyen D. Edward

Department of Construction Management and Quantity Surveying, Faculty of Engineering and the Built Environment, University of Johannesburg, Johannesburg, South Africa
caigbavboa@uj.ac.za, emayok@gmail.com, davidlooyen@gmail.com

Abstract. The construction industry is key to the development and economic growth of any country, either developing or developed. The role of the industry in the achievement of infrastructures that are economically, socially and environmentally sustainable cannot be over-emphasized. It is therefore necessary for construction industry to continuously pursue sustainable construction projects. This study examines factors influencing the adoption and implementation of sustainable construction goals as well as the impact of facility management in achieving and sustaining the practice in the construction industry. Data were collected through well-structured questionnaires administered on relevant stakeholders that are involved in various activities in the construction industry. The policies and actions which affects the transition of companies towards sustainable construction practices include implementation of sustainable policies in the design stage, waste reduction practices, effective reuse and recycling of materials, utilization of life cycle costing and implementation of facility management principles. To improve sustainable practices in the industry, concerned stakeholders must take into consideration maintenance and running of all facilities, implementation of life cycle costing, training of local construction companies on sustainable methodologies and creating awareness within the industry on benefits associated with sustainable construction. This study has contributed to the body of knowledge by examining the role of facility management in the achievement of sustainable construction, it is therefore recommended that the identified policies and techniques as well as highlighted facility management processes are given consideration by construction stakeholders in their quest to improving performance of construction projects.

Keywords: Construction project · Facility management · Green building · Life cycle cost · Sustainable construction

1 Introduction

The call and clamor for sustainability has become a global issue and has been on the increase over the years. The recent modification of the United Nations' Millennium Development Goals (MDG) to Sustainable Development Goals (SDG) is a pointer that the demand for such projects will continue to rise. With the depletion of natural

J. Calautit et al. (eds.), *Towards Sustainable Cities in Asia and the Middle East*, Sustainable Civil Infrastructures, DOI 10.1007/978-3-319-61645-2_3

resources fast becoming a problem, national and international governments are under pressure to mitigate these issues through sustainable practices. In the construction industry, this implies that informed clients and owners of construction projects will continuously demand for complex but sustainable projects that reduce energy, safe cost, beautify the environment, socially safe, etc. Khalil *et al.* (2011) defined sustainable development as a form of project that meets the needs of the present without compromising the ability of future generations. Adewunmi *et al.* (2012) described sustainable development as the line between environmental, economic, and social sustainability and the three factors are necessary for any true measurement of sustainability.

The possibility to supply shelter, schools, water and energy systems, industry, communications, commercial or leisure facilities, alongside jobs and revenue generated make construction one of the main activities worldwide (Pietrosemoli and Monroy 2013). Rashid and Yusoff (2015) stated that energy is consumed directly during building construction. It was further highlighted that recent studies have shown that buildings around the world are responsible for 30–40% of energy use and 40–50% of world greenhouse gas emissions. Although the building industry is crucial for social and economic development, the environmental impacts of the process are also important.

Zeiler *et al.* (2012) informed that sustainability is a crucial issue for our future and professionals, especially architects and engineers involved in the design process, have an important influence on the practice. The same is also important for facility managers in the reduction of the built environment impact on the natural resources. Due to increasing environmental awareness and legislative pressures, facility management as a profession is required to play an active role in the environment. According to Hodges (2005), the primary incentive for the involvement of facility managers in sustainable and green building practices is to reduce energy consumption and the subsequent reduction in the reliance on fossil fuel to produce that energy. However, Adewunmi *et al.* (2012) indicated that there is a lack of literature on integrating sustainability into facility management in developing countries. Therefore, the purpose of this study is to examine the benefits of facility management principles in the achievement of sustainable construction goals and practices in South African construction industry with a view to justify the inclusion of the practice in the design process of construction activities.

2 Review of Literature

2.1 Sustainability in Construction

The construction industry contributes significantly to the socio-economic development and also consumes a large amount of energy and natural resources (Asif *et al.* 2007). It was further noted that the industry do not only consumes 40% of materials entering the global economy but it also generates 40–50% of the global greenhouse gasses. However, Oke *et al.* (2015) concluded that the industry is currently moving towards social, economic and environmental sustainable construction. Social or societal sustainability aims at providing healthy and liveable communities while Economic or financial sustainability is concerned with the long-term financial benefits associated with the practice. Environmental, that is, physical sustainability, refers the preservation of the natural environment, the conservation of non-renewable resources, the mitigation of waste and the implementation of construction practices which aim at sustaining the environment. Almahmoud and Doloi (2012) concluded that sustainability development stipulates that society, the economy and the environment are interrelated.

In a report by Pacific Northwest National Laboratory (PNNL) that highlighted the results of a post occupancy evaluation of 22 green federal buildings from across the United States, Zeiler *et al.* (2012) noted that, on the average, green buildings, compared to commercial buildings uses 25% less energy, emit 34% less carbon dioxide, cost 19% less to maintain and have 27% more satisfied occupants. These results show that organisations that are willing to adopt sustainable practices will not only strides towards preserving the environment but also benefit in terms of finances. With less consumption of energy, lower maintenance costs and more satisfied occupants, the organisation will be able to save costs and satisfied occupants will result in higher productivity for the society.

The US Green Building Council in energy and environmental design has implemented a strategy whereby organizations may strive towards receiving certification indicating that the particular organization has addressed environmental impact of its facilities Presley and Meade (2010). The incentives on offered within the UK construction industry to promote innovation towards sustainable construction practices range from financial incentives such as higher rental rates or property tax rebates to reputation in the community. It was stated that companies shall receive incentives for their ability to initiate savings through introduction of sustainable construction practices.

2.2 Facility Management and Sustainable Construction

McCauley, *et al.* (2012) observed that, with the realization of the need for a more sustainable future and the need for alternative economic solutions, have shown that traditional construction methods need to be rectified with the inclusion of more innovative approaches towards construction methods. One of the best ways to achieve these goals is through the implementation of facility management principles, which will assist with providing a new focus for the study of buildings. Due to the facility manager's vast experience in the operation of a facility during the post-construction phase, the facility manager will be able to identify possible complications which may arise due to poor decision making in the design stage. Therefore by involving a facility manager in the design process, Ogungbile and Oke (2015) concluded that an organization can ensure that all decisions will be made with future implications taken into account. McCauley *et al.* (2012) concluded that the introduction of facility manager at the beginning of a projects life cycle, has the potential to significantly increase sustainability and at the same time promote better construction practices.

The willingness and ability of the facility management and relevant stake holders, within an organization, to adopt greater environmental awareness and responsibility is critical in striving towards a higher degree of sustainability in the construction and development sector. The primary objective of the facility manager is to oversee the maintenance and upgrading of the built environment within a specific organization.

In view of this, a facility manager needs to integrate the employees of the organization with the subtleties of the infrastructure on a single incorporated platform. The duties of the facility manager is directly affected by the changes in the sustainability area which have been altered by climate change (Khalil *et al.* 2011). A facility manager needs to be able and willing to incorporate sustainability into their thought processes and decision making. Strategic planning for the future is one of the facility manager's primary objectives; therefore, the ability to incorporate sustainability into the future plans of an organization can be greatly influenced by the facility management team. Life cycle performance cost calculations is an important decision support tool for sustainability. Zeiler *et al.* (2012) noted that facility managers involved in the design process are able to implement the life cycle performance costing tool in order to determine the most cost effective sustainable solution, with the entire life cycle of the project taken into consideration.

Khalil *et al.* (2011) noted that with cost effective and environmental issues being the main concern in every field of economic activity, facility managers will need to adapt accordingly in the future. Organizations worldwide have identified the need to implement sustainable construction, therefore they will need to initiate the incorporation of facility management principles into their decision making processes. Ogungbile and Oke (2015) insisted that the responsibilities of the facility manager has expanded to accommodate green procurement and energy management, this implies that the management of sustainability and reduction of carbon emissions has also been incorporated into the responsibilities of the facility manager. The scope of work of the facility manager is expanding beyond just maintaining a facility. It is due to such changes in responsibility that facility management principles should be introduced within the design stage of a project. International Facility Management Association (2007) aims to assists facility management professionals in planning and implementing strategic sustainability programs. The integration of sustainability within the facility management process creates a relationship which enables sustainability concepts and facility management principles to work hand in hand. More so, facility management professionals have long been concerned with and engaged in the environmental aspects of the facilities they operate, especially from the perspective of energy conservation and high-performance buildings. Facility managers are well equipped to improve the sustainability of the organization for which they work, organizations however need to be convinced that sustainable buildings will benefit them in the future.

3 Research Methodology

Using survey design, factors and variables that are relevant to the study were elicited from existing literature materials. This is to gather responses pertaining to the principles underlying facility management in the quest of adopting them for sustainable construction. The population consist of professionals within the various tiers of construction industry, that is, building, civil and industrial engineering as well as contractors, clients, owners and financiers of construction projects. These includes architects, quantity surveyors, engineers, project managers, construction managers, facility managers, contractors and agents of public and private clients. The register of the professionals were obtained from their various professional bodies while the list of contractors were retrieved from Construction Industry Development Board (CIDB).

Closed-ended questionnaires arranged in sections in respect to the objectives of the study were adopted to solicit needed information from respondents. Content validity was enhanced by considering information in existing literature materials in the areas of sustainability and facility management. Prior to the actual distribution, a pilot study was conducted to assess the suitability and correctness of the instrument. Various comments, suggestions and observations pointed out by the respondents were examined and effected in the final document.

The respondents were not influenced in any way as to how to complete the questionnaire but necessary information were provided on the cover page of the instrument. The necessary guide and instructions as well as expected duration to complete the instrument were provided. The research questions were arranged in a logical order and were constructed to provide opinions by marking one choice answer on a 5-point Likert scale. In this way, it easier to present the data in a logical sequence which allow for authoritative analysis of the data. Based on the adopted scale, mean item score (MIS) and standard deviation (SD) were calculated and employed to determine the significance and importance of the identified variables and factors. By corroborating the results of analyzed data with existing information in the literature, emerging findings were strengthened and recommendations pertaining to sustainable facility management principles were posited.

4 Findings and Discussion

During the period of the survey, 98 questionnaires were administered on the respondents while 82 were retrieved. 80 of them were finally used for analysis because the remaining 2 of the retrieved instruments were not correctly and completely filled.

4.1 Respondents' Information

The respondents' professional qualification indicates that about 75% are construction professionals while the remaining 25% are contractors, clients' agents and other stakeholders in the construction industry. The others includes professionals such as town planner, architectural technician, interior designer, executive secretary/buyer, graphic designer, project fundamentals specialist, etc. With an average of about 7 years experience in the construction industry, 15% of the respondents are engaged in the public sector, 40% in the private sector and 45% provide their services for both sectors. They have been involved in various kinds of building, civil and industrial engineering projects including renovations, housing, office complex, airport projects, etc.

4.2 Influences of Sustainable Construction

Table 1 portrays the respondents' ranking of company policies and actions that affect the transition towards a sustainable construction industry. It reveals that

Table 1. Policies and actions affecting sustainable construction

Company policies and actions	$\bar{x}$	σX	R
Implement sustainable policies in design stage	4.13	8.76	1
Waste reduction practices	4.08	7.16	2
Effective reuse and recycle of materials	4.03	11.05	3
Company emphasizes the importance of preserving the environment	4.00	7.62	4
Utilization of whole life costing	4.00	7.39	4
Utilization life cycle costing	3.98	9.83	5
Sustainable construction considered in the design stage	3.98	7.52	5
Consideration of economic sustainability	3.95	10.39	6
Consideration of environmental sustainability	3.95	9.66	6
Future considerations are made in the design stage	3.95	8.57	6
Environmental protection workshops/training	3.93	8.37	7
Conduct environmental assessments	3.93	10.98	7
Knowledge of facility management principles	3.93	7.26	7
The company treats all facilities as assets	3.90	7.12	8
Implementation of facility management principles	3.90	9.52	8
Sustainable construction workshops/training	3.88	6.16	9
Involvement of facility managers in the design stage	3.88	7.65	9
Environmental protection intentions	3.85	10.86	10
Utilization of the facility management profession	3.83	8.29	11
Sustainable construction intentions	3.78	12.03	12
Consideration of Non-renewable resources preservation	3.73	8.09	13
Green procurement strategies	3.73	8.28	13
Consideration of social sustainability	3.55	9.38	14

$\bar{x}$ = Mean item score; σX = Standard deviation; R = Rank

implementation of sustainable policies in the design stage was ranked first with a mean score of 4.13 and standard deviation of 8.76. This implies that the most important policies and actions affecting the transition of companies towards sustainable construction practices in the study area are related to implementing sustainable policies in the design stage. Other important factors are waste reduction practices, effective reuse and recycling of materials, company emphasis on the importance of preserving the environment, utilisation of whole life costing, utilisation of life cycle costing, sustainable construction considered in the design stage, consideration of economic sustainability, consideration of environmental sustainability, future considerations are made in the design stage, environmental protection workshops/training, conduct environmental assessments, knowledge of facility management principles, the company treats all facilities as assets and implementation of facility management principles. The least influential factors with mean item score of 3.88 of 5.00, are concerned with involvement of facility managers at the design stage and organizing trainings and workshops on sustainable construction.

These findings were reinforced by the study of McAuley *et al.* (2012), it was concluded that the involvement of facility manager at the beginning of a projects life cycle, has the potential to significantly increase sustainability and at the same time promote better construction practices. The findings agree with the study of Elmualim *et al.* (2012), it was noted that governments and international bodies are implementing legislations and regulations regarding efficient energy management, removal of waste and the reduction of carbon emissions in order to achieve sustainability goals. Presley and Meade (2010) stated that it is important for companies to install managerial tools and methodologies with which to monitor and improve their performance as they move towards sustainable construction and design. The results further reinforced the study of Khalil *et al.* (2011) where it was pointed that the willingness and ability of the facility management and relevant stake holders, within an organization, to adopt greater environmental awareness and responsibility is critical in striving towards a higher degree of sustainability in the construction and development sector.

The results on whole life costing also agree with the study of South African Council for the Quantity Surveying Profession (SACQSP 2014a), it was noted that the practice takes into consideration all relevant and agreed cost flows over a period of time to determine an economic value. In this case, the relevant costs are those needed to achieve defined levels of performance, including reliability, safety and availability. Hodges (2005) pointed that life cycle costing aims to account for the costs associated with all the building stages and reduce those costs to present value. More so, South African Council for the Quantity Surveying Profession (SACQSP 2014b) observed that with life cycle costing and whole life costing, the facility manager will be able to reduce overhead costs for an organization. The results also concur with the assertion of Meistad and Valen (2012) that there are a variety of suggestions on how knowledge of operation, cleaning and maintenance can be taken into consideration at an early stage in construction projects in order to ensure effective planning takes place. The results further agree with the observation of Hodges (2005) that sustainable practices are able to generate enormous economic benefits for organizations that are willing to explore and implement the practices. According to Zeiler *et al.* (2012), sustainability assessment tools are currently implemented in important decisions regarding the construction industry by taking sustainability into account. This is in agreement with William and Sutrisna (2010) conclusion that the main aims of sustainability within the construction industry is to reduce the consumption of natural resources such as energy and water, develop sustainable building practices and designs, increase recycling and reuse of recycled materials, preserve the surrounding ecosystem and reduce CO_2 emissions.

4.3 Facility Management as Drivers of Sustainable Construction

Table 2 reveals the respondents ranking of facility management processes that can be initiated in order to improve sustainable construction practices within the South African construction industry. The table shows that the most important factor with a mean score of 4.35 and standard deviation of 9.07 is that design teams must take into consideration post-construction occupancy, maintenance and running of a facility. Other important facility management processes include design teams initiating the implementation of life cycle costing and whole life costing, training of local construction companies on sustainable methodologies, need for companies to be made aware of environmental,

Table 2. Processes to improve sustainable construction practices

Facility management processes	x̄	σX	R
Design teams must take into consideration post construction occupancy, maintenance and running of a facility	4.35	9.07	1
Design teams must initiate the implementation of life cycle costing and whole life costing	4.28	8.62	2
Training of local construction companies on sustainable methodologies	4.15	11.37	3
Companies need to be made aware of the environmental, social and economic benefits associated with sustainable construction	4.15	9.71	3
Engage facility managers in the full life cycle of the project	4.13	8.74	4
Ensuring employees are equipped and able to use assessment tools in order to measure their impact on the environment	4.13	8.74	4
Sustainable construction practices long term benefits outweigh that of traditional building practices	4.05	8.97	5
Facility managers must be included within the design stage	4.03	6.06	6
Educating employees as to the importance of sustainable practices	4.03	10.12	6
Implement training strategies in order to create environmental protection awareness	3.95	9.06	7
Incentives are available to construction companies who implement sustainable construction practices	3.93	5.48	8

x̄ = Mean item score; σX = Standard deviation; R = Rank

social and economic benefits associated with sustainable construction, engaging facility managers in the full life cycle of construction project, ensuring employees are equipped and able to use assessment tools in order to measure their impact on the environment, including facility managers in the design stage, educating employees as to the importance of sustainable practices, implement training strategies in order to create environmental protection awareness, and making incentives available to construction companies who implement sustainable construction practices.

In agreement with the findings of the study, Meistad and Valen (2012) concluded that knowledge of operation, cleaning and maintenance should be taken into consideration at an early stage of construction projects in order to ensure effective planning. More so, Hodges (2005) posited that for sustainability and green building to succeed, the facility manager will need to build an economic case by implementing life cycle costing and whole life costing systems. The results of this study agree with the assertion of Almahmoud and Doloi (2012) that social sustainability is achieved by engaging with employees, local communities, clients and other stakeholders in the community.

On the involvement of facility managers, McCauley *et al.* (2012) stated that there is a belief amongst facility managers, project managers and architects that the facility manager can play a key role in enhancing sustainability potential of construction a projects. However, Presley and Meade (2010) noted that it is important for companies to install managerial tools and methodologies with which to monitor and improve their performance as they move towards sustainable construction and design. This is because sustainable practices are able to generate enormous economic benefits for organizations that are willing to explore and implement sustainable practices (Hodges 2005). The findings on incentives agree with the recommendation of RICS 2004 that incentives need to be offered to companies that promote innovation towards sustainable construction practices.

5 Conclusion and Recommendations

The study has explored various company policies and actions affecting the transition towards sustainable construction practices as well as facility management principles which can be initiated in order to improve sustainable construction practices. The negative effects of the industry and the immense consumption of non-renewable resources necessitate the need for the industry to adopt sustainability practices and technique. The implementation of facility management principles is able to contribute immensely towards reducing these negative effects and assist in the achievement of sustainable construction.

The factors affecting implementation of sustainable construction practices are concerned with the implementation of sustainable policies in the design stage and waste reduction practices. These are related to effective reuse and recycling of materials, utilisation of whole life costing, environmental protection workshops/training as well as knowledge of facility management principles. In order to improve sustainable practices within the construction industry, various facility management processes that could be initiated were also highlighted. These include consideration of post-construction occupancy, maintenance and running of facility, implementation of life cycle costing and training of local construction companies on sustainable methodologies and their importance.

In view of the highlighted policies for implementing sustainable construction projects, clients, owners, design team and other stakeholders involved in construction of projects should focus on achieving social, environmental and economic sustainability equally in order to ensure that the triple bottom line is addressed accordingly. More so, the design teams should take into consideration post-construction occupancy, maintenance and operational costs when designing new facilities. It is also recommended that design teams develop systems whereby alternate methods and decisions are scrutinised by through life cycle costing or whole life costing techniques prior to actual construction phase of projects.

References

Adewunmi, Y., Omirin, M., Koleoso, H.: Developing a sustainable approach to corporate FM in Nigeria. Facilities **30**(9), 350–373 (2012)

Almahmoud, E.S., Doloi, H.K.: Social sustainability health check: a model for integrating stakeholders' interests in evaluating and optimising social sustainability, performance of construction projects. In: Proceedings of CIB W070, W092 and TG72 Joint Conference on Procurement Systems, Facilities Management and Maintenance and Public Private Partnership: Delivering Value to the Community, pp. 98–105. University of Cape Town (2012)

Asif, M., Muneer, T., Kelley, R.: Life cycle assessment: a case study of a dwelling home in Scotland. Build. Environ. **42**(3), 1391–1394 (2007)

Elmualim, A., Valle, R., Kwawu, W.: Discerning policy and drivers for sustainable facilities management practice. Int. J. Sustain. Built Environ. **1**(1), 16–25 (2012)

Hodges, C.P.: A facility manager's approach to sustainability. J. Facil. Manag. **3**(4), 312–324 (2005)

International Facility Management Association: Facility Management Forecast 2007: Exploring the Current Trends and Future Outlook for Facility Management Professionals. International Facility Management Association, Houston (2007)

Khalil, N., Husin, H.N., Mahat, N., Nasir, N.: Sustainable environment: issues and solutions from the perspective of facility managers. Procedia Eng. **20**, 458–465 (2011)

McAuley, B., Hore, A.V., West, R., Wall, J.: The economic case for early adoption of facilities management. In: Proceedings of CIB W092/W070/TG72 Joint Conference on Procurement Systems, Facilities Management and Maintenance and Public Private Partnership: Delivering Value to the Community, pp. 215–223. University of Cape Town, South Africa, 23–25 January 2012

Meistad, T., Valen, M.S.: Adding value and sustainability by involving facility managers in design phase. A preliminary study of Norwegian pilot projects of energy efficient buildings. In: Proceedings of CIB W092/W070/TG72 Joint Conference on Procurement Systems, Facilities Management and Maintenance and Public Private Partnership: Delivering Value to the Community, pp. 345–352. University of Cape Town, South Africa, 23–25 January, 2012

Ogungbile, A.J., Oke, A.E.: Assessment of facility management practices in public and private buildings in Akure and Ibadan cities, South Western Nigeria. J. Facil. Manag. **13**(4), 366–390 (2015)

Oke, A.E., Aghimien, D.O., Olatunji, S.O.: Implementation of value management as an economic sustainability tool for building construction in Nigeria. Int. J. Managing Value Supply Chains **6**(4), 55–64 (2015)

Pietrosemoli, L., Monroy, C.R.: The impact of sustainable construction and knowledge management on sustainability goals. A review of the Venezuelan renewable energy sector. Renew. Sustain. Energy Rev. **27**, 683–691 (2013)

Presley, A., Meade, L.: Benchmarking for sustainability: an application to the sustainable construction industry. Benchmarking: Int. J. **17**(3), 435–451 (2010)

Rashid, A.F.A., Yusoff, S.: A review of life cycle assessment method for building industry. Renew. Sustain. Energy Rev. **45**, 244–248 (2015)

RICS: Sustainability, the built environment – an agenda for action. R. Inst. Charted Surveyors, London (2004)

South African Council for the Quantity Surveying Profession: Professional skills module number 11: Undertake whole life appraisals of built environment projects. Quantity Surveying Profession Act 49 of 2000 (2014a)

South African Council for the Quantity Surveying Profession: Professional skills module number 18: Research methodology and report writing. Quantity Surveying Profession Act 49 of 2000 (2014b)

Williams, D., Sutrisna, M.: An evaluation of the role of facilities managers in managing sustainability and the remedial actions in reducing CO2 emissions in the built environment. In: Proceedings of COBRA 2010, Paris, France, 2–3 September 2010 (2010)

Zeiler, W., Maaijen, R., Maassen, W.: Decision support for facilities management of the future: sustainability accelerator. In: Proceedings of CIB W070, W092 and TG72 Joint Conference on Procurement Systems, Facilities Management and Maintenance and Public Private Partnership: Delivering Value to the Community, pp. 353–360. University of Cape Town (2012)

Implementing Tilt-Up Method for Sustainable Construction

Ayodeji E. Oke(✉), Clinton O. Aigbavboa, and Yongama L. Thole

Department of Construction Management and Quantity Surveying,
Faculty of Engineering and the Built Environment,
University of Johannesburg, Johannesburg, South Africa
emayok@gmail.com, caigbavboa@uj.ac.za,
yongamathole@yahoo.com

Abstract. Construction projects are becoming more complex and clients are increasingly demanding innovative and sustainable projects. These have resulted in several challenges and limitations with the use of traditional construction techniques and as a result, there has been a paradigm shift to technology driven method known as tilt-up construction system. This study examined the viability and benefits of this method as well as their perceived effect on activities of workmen in the South African construction industry in the quest to adopting as a tool for sustainable construction. Based on information obtained from existing literature on the subject area, questionnaires were designed and administered on construction professionals and other stakeholders with knowledge of tilt-up construction method. Structural integrity, clients' satisfaction as well as cost, quality and time will be enhanced by application of the method thereby enhancing sustainable construction. This will be achieved through minimization of waste, time delays and construction site traffic. Against the negative notion that the practice will lead to reduced workmen on construction sites, tilt-up method will help to minimize human error, site accidents and greater use of machinery. Stakeholders in the construction industry will benefit from the result of this study in their quest to promoting sustainable developments. There is therefore the need for adequate tilt-up training of personnel and professionals involved in construction activities for smooth transition from conventional method of construction. Legislation for the purpose should also be provided by government and appropriate incentives should be provided to encourage the usage of the system.

Keywords: Conventional construction method · Construction process · Construction resources · Housing · Sustainable construction · Tilt-up

1 Introduction

The nature of construction project delivery has changed greatly in the recent past with more project owners demanding shorter time delivery of construction projects and an early indication of the final project cost at the start of the project. Hallinan and Guan (2006) noted that tilt-up construction method has the capacity to address concerns relating to safety of people, durability of structures in withstanding high pressures of weather as well as cost, time and quality performance of construction projects.

J. Calautit et al. (eds.), *Towards Sustainable Cities in Asia and the Middle East*,
Sustainable Civil Infrastructures, DOI 10.1007/978-3-319-61645-2_4

The system has been in used for some years in some developed countries such as United States of America, Australia, United Kingdom with outstanding results in all areas that normally cause concern. These countries have mastered the Tilt-up method and as a result they are heading into using it as the norm form of construction. It is also worthy of note that some developing countries, including South Africa has also adopted the system but level of awareness and usage is still at the lower ebb. According to Tilt-up Concrete Association (TCA 2011), the system offers a more economical and durable solution than competing conventional and traditional building methods. The inception of the use of Tilt-up systems has been around for almost two centuries and its development and growth has allowed massive buildings which would have otherwise taken time and tremendous effort if conventional methods were used (Harris 2012). For that reason this research has been motivated by Tilt-up's technological advancement and its performance in the building of the Gateway Shopping Mall in Umhlanga Durban Kwa Zulu Natal. According to Tilt-up Concrete Association (TCA 2015), it is the largest Tilt building in square meters in the world. In view of this, this study examines the viability and benefits of tilt-up construction method in the achievement of sustainable construction goals.

2 Tilt-Up Method and Sustainable Construction

Achieving sustainable projects has become a new direction for the construction industry with increase in the use of technology driven method rather than traditional or conventional techniques (Oke et al. 2015). Hodges (2005) insisted that sustainability is an important social and economic issue. The environment cannot be disregarded while excessive consumption of natural resources is taking place, same is applicable to economic and social well-being of people. The social, economic and environmental impacts of sustainability have become well known as the triple bottom line. The triple bottom line can be used to improve the awareness of sustainability within organizations as well has to highlight the economic benefits of the better treatment of the environment.

The South African construction industry has always been a labor intensive with the goal of aiding community development through increase in unemployment rate. Sometimes, the percentage of residing community members to be employed are stipulated in government projects with a view to give back to the community and develop the skills of dwellers (Hanscomb 2004). Tilt up in general is a very advance construction method which uses the least time and labor when compared to other construction methods. This is so as construction companies aim to be as profitable as possible while completing projects at the shortest amount of time (Kitching 2014). Hagerstown (2012) observed that the cost of using machinery such as a tractor, loader and backatter in tilt-up system is cheaper than hiring workmen to use wheel barrows to load aggregate to mixing bay and pouring same.

3 Research Methodology

Variables and factors relating to tilt-up construction methods were derived from existing literature and various areas were reviewed accordingly. Using the set of variables, quantitative method was adopted in the collection of data for the study through survey of appropriate individuals. The target group were quantity surveyors, architects, engineers, construction managers as well as construction stakeholders with experience and knowledge of the use of the tilt-up construction system.

Closed-ended questionnaire prepared in official language of the country, that is, English, was adopted as the research instrument for the study. At the initial stage, a list of potential respondents were drafted and register of construction professionals were obtained from various professional bodies within the study area. Email, post and hand delivery were adopted for the distribution of the research instrument and they were retrieved through the same means. A questionnaire took an average duration of about 10 min to complete and all respondent were given the freedom to phone and acquire information about questions they do not understand. Anonymity and confidentiality was maintained throughout the study and this was achieved by eliminating questions that could link or identify respondents.

Background questions in the first section of the questionnaire was used as the basis for determining the ability of respondents to provide adequate information regarding the study. A cover letter was also included to provide necessary information concerning the study and instruction needed to complete the document. The second section of the questionnaire was structured to address the main objectives of the study and a 5-point Likert scale was adopted to solicit require information. The adopted scale is as follow: 1 = Strongly disagree; 2 = Disagree; 3 = Neutral; 4 = Agree; and 5 = Strongly agree. The 5-point scale was converted to mean item scores (MIS) and standard deviation (SD) for each of the identified variables. The outcome of the MIS was then used to identify or rather determine the rank of each item.

4 Findings and Discussion

Out of the 70 questionnaires circulated to gather information for the study, 44 were retrieved while 40 were completely and correctly filled and certified fit for further analysis.

4.1 Respondents' Information

The analyzed data indicates that respondents for the study have an average of about 8 years of experience in the construction industry. Findings relating to their knowledge of tilt-up construction system reveals that about 23% have average knowledge, 36% have sufficient knowledge while the remaining 41 have a vast knowledge of the method. More so, 50% of the respondents are from small and medium enterprises, 23% are from grade 6 and grade 8 companies while 27% are from grade 9 company as rated by Constriction industry Building Development (CIDB). This implies that respondents are widely spread across various levels and types of construction companies with basic knowledge of tilt-up construction system indicating a good tendency of obtaining adequate and balanced responses.

4.2 Advantages and Viability of Tilt-Up Construction Method

Findings relating to stand out advantages of tilt-up construction method are indicated in Fig. 1. The most important is structural integrity with 36%, followed by time effectiveness, minimal maintenance costs, clients' satisfaction, improve quality of works and effective cost.

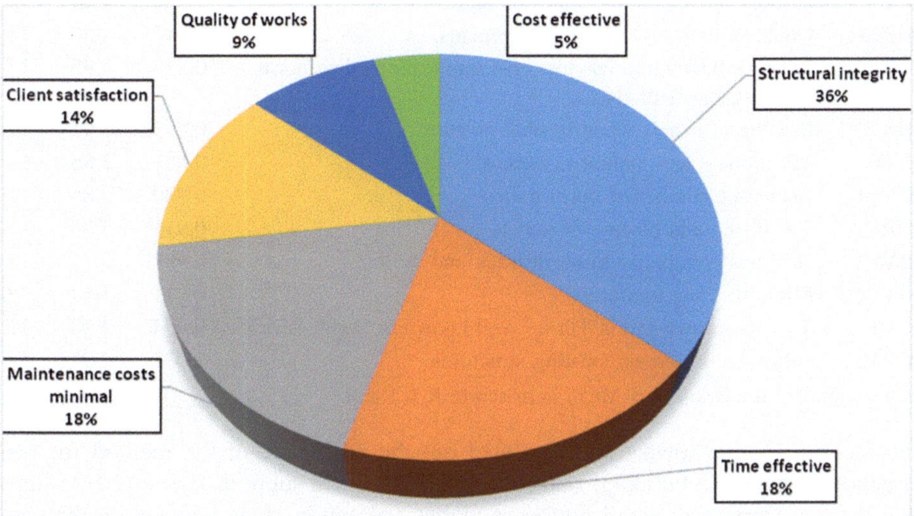

Fig. 1. Advantages of tilt-up construction method

Table 1 shows that the ranking of viability of Tilt-up construction method in the South African construction industry in the quest to achieving sustainable construction goal. It reveals that the most important factors is minimization of waste with a mean score of 4.22 and standard deviation of 0.899. The method will also minimize time delays; easy to adapt to high rise and RDP construction; increase client satisfaction; easy to adapt to norm house construction; enhance quality standards; reduce site traffic; aid faster assembling of roof components; and has less cost implications than conventional and traditional construction methods. The average importance of the method to the construction industry includes use of less natural resources, minimize site traffic and its ability to be used for existing structures.

These findings are similar to Hallinan and Guan (2006) where it was stated that the accurateness of the Tilt-up construction method aids elimination of waste as materials are poured into provided casting space. More so, the minimization of working trades eliminates labour waste thereby, maximizes value for money and aid better client

Table 1. The viability of tilt-up in South African construction

	Viability	σX	x̄	R
CV4	Tilt-up construction minimizes waste	0.899	4.22	1
CV5	Tilt-up construction minimizes time delays	0.919	4.22	2
CV9	Easy to adapt to high rise building construction	0.964	4.22	3
HRC4	Easy to adapt to RDP construction	1.072	4.17	4
HRC1	The casting of the panels	1.022	4.13	5
CV2	Increases client satisfaction	0.879	4.08	6
CV8	Easy to adapt to norm house construction	0.880	4.08	7
CV11	Exceeds current quality standards	0.877	4.08	8
HRC5	Minimized site traffic	0.891	4.08	9
CV7	Regular movement of cranes on site	1.052	4.04	10
HRC6	Quick assembling of roof components	1.012	4.00	11
CV3	Tilt-up construction has less cost implications than norm construction methods	0.990	3.95	12
HRC3	Bracing of panels when in final position	0.893	3.91	13
CV6	Tilt-up uses less natural resources	0.888	3.86	14
CV14	Decreases number of norm trades	0.880	3.86	15
HRC2	The lifting and placing of panels	0.911	3.86	16
CV13	Holistically superior to norm brick and mortar	0.988	3.73	17
CV12	Minimizes site traffic	0.971	3.60	18
CV1	To your knowledge is Tilt-up well known in South Africa	0.954	3.20	19
CV10	Adaptable to use on existing structures	1.032	2.97	20

σX = Standard deviation; x̄ = Mean item score; R = Rank

satisfaction. Haggerstown (2012) pointed out the advantage of the method for construction of high rise buildings and low cost housing. In support, Harris (2013) stated that the construction of small houses is viable and can be done keeping all the characteristics and advantages of Tilt-up. Kitching (2014) highlighted the benefits of the method in respect to cost implication. It was stated that the system is cheaper for masonry work and it is proven to be one of the most cost and quality effective construction method.

4.3 Effect of Tilt-Up Method on Construction Labor Force

Table 2 shows construction workforce issues that will be affected by the implementation of tilt-up construction system. The major effect is that it will minimize human errors and site accidents. Other important benefits, with mean item score of 3.70 and above are that the system will promote greater use of machinery, less site safety personal will be required on site; and it will promote the upgrading of professional skills. The least issues are related to the perceived disadvantages of the system if fully implemented. These included reduction in demand for certain trades; harming of trade training institutes; higher levels of skills will be required; minimal need for unskilled labor and brick layers will become obsolete with a mean score of 2.22 and standard deviation of 0.879.

Table 2. Effects of tilt-up method on labor force

	Effects	σX	x̄	R
LB6	Tilt-up minimizes human error	1.033	3.95	1
LB7	Tilt-up minimizes site accidents	1.042	3.91	2
LB9	Promotes greater use of machinery	1.036	3.86	3
LB8	Less site safety personal is needed	0.987	3.77	4
LB10	Promotes the upgrading of professional skills	0.987	3.77	5
LB5	Implementation will create a demand for certain trades	0.987	3.6	6
LB4	Implementation will harm trade training institutes	1.056	3.2	7
LB2	Implementation will require higher levels of skills	0.897	2.84	8
LB1	There is minimal need for unskilled labor	0.983	2.8	9
LB3	Brick layers will become obsolete	0.879	2.22	10

σX = Standard deviation; x̄ = Mean item score; R = Rank

These results are similar to the findings of Hagerstown (2012); Harris (2013). It was stated that the use of Tilt-up construction does limit the use of intensive labor force and it is only needed at specific phases or processes of construction such as digging of trenches, moving of concrete, laying of reinforcement and placing of braces on tilt walls. This implies that Tilt-up construction method was devised to employ and train labor instead of eliminating them. Tilt-up Concrete Association (TCA 2013) also observed that Tilt-up caters for client's needs, make construction sense and architects have more freedom of design without any compromise to structural integrity and workforce activities.

5 Conclusion and Recommendations

The study expanded the body of knowledge about tilt-up method in the construction industry with emphasis on its usefulness as sustainable construction tool as well as its general effect on activities of construction workforce. The major benefits of adopting tilt-up system in the construction industry is to minimize cost, time and materials waste, minimizes site traffic, as well for easier, safer and faster construction of high rise buildings and RDP houses. For workmen, the method will minimize human error, reduce site accidents and promote the development of their professional skills.

Public clients, that is, government agencies and parastatals at all levels should implement element of tilt-up in public tenders especially for schools, clinics and low cost housing projects for sustainable benefits relating to economic, social and the environment. Professionals in the construction industry should equip themselves with necessary knowledge relating to principles and application of tilt-up construction method so as to be able to advise clients on the adoption of the method as a feasible sustainable construction methods. It is also essential for construction companies to organize training for their workmen and personnel on the essential of tilt-up method in their quest for the delivery of sustainable projects to the satisfaction of their clients.

References

Hagerstown, M.D.: Pre-engineered steel structure vs tilt-up concrete construction, FedEx Ground Distribution Hub, pp. 49–58 (2012). http://www.engr.psu.edu/. Accessed 08 Nov 2015

Hallinan, P., Guan, H.: Layered finite element analysis of one-way and two way concrete walls with openings. Adv. Struct. Eng. – Int. J. **10**, 55–72 (2006)

Hanscomb, M.P.: Design and build becoming a revolution. Int. Constr. Intell. **16**(6), 1–6 (2004)

Harris, A.H.: Tilt up engineering (2013). http://tilt-up.org/. Accessed 08 Nov 2015

Harris, C.: Concrete tilt-up construction for sustainable world (2012). https://www.vrmca.com/. Accessed 08 Nov 2015

Hodges, C.P.: A facility manager's approach to sustainability. J. Facil. Manag. **3**(4), 312–324 (2005)

Kitching, D.: Tilt-up construction. Bedrock concrete construction specialist. http://www.bedrockgroup.co.za. Accessed 08 Nov 2015

Oke, A.E., Aghimien, D.O., Olatunji, S.O.: Implementation of value management as an economic sustainability tool for building construction in Nigeria. Int. J. Manag. Value Supply Chains **6**(4), 55–64 (2015)

Tilt-up Concrete Association: The construction of tilt-up (2011). http://tilt-up.org/. Accessed 08 Nov 2015

Tilt-up Concrete Association: TCA safety guideline for tilt-up concrete construction. Safety committee of the tilt-up concrete association, **13**(1) (2013). http://tilt-up.org/. Accessed 08 Nov 2015

Tilt-up Concrete Association: TCA safety guideline specifications for site-cast tilt-up concrete construction (2015). http://tilt-up.org/. Accessed 12 June 2016

"Mission for the Sanitary India: A Case Study of Aligarh City" Uttar Pradesh, India

Harit Priyadarshi[1(✉)], Sangharsh Rao[2], Shahbaz Shamshad Khan[3], and Divyanshu Vats[3]

[1] Department of Civil Engineering, Mangalayatan University, Beswan, Aligarh 202145, UP, India
harit.priyadarshi@mangalayatan.edu.in
[2] Remote Sensing Applications Centre, Jankipuram, Kursi Road, Lucknow 226021, UP, India
[3] Diploma in Civil Engineering, Mangalayatan University, Beswan, Aligarh 202145, UP, India

Abstract. India is a religious country with ancient civilization and having population diverse religion. Our country having so much pride in unity in diversity lacks behind in meeting. International standards of sanitation and cleanliness realizing poor sanitation. The Father of nation Mahatma Gandhi had a dream of clean India. Gandhi played a great role for making Indian clean and green. Moreover the Government of India has followed the foot prints of Gandhi and took his dream in positive way by starting a Mission for the sanitary India. The Mission was started by Honourable Prime Minister of India Mr. Narendra Damodardas Modi on 2nd of October, 2014 in the honour of Mahatma Gandhi 145th Birth anniversary at Rajghat, New Delhi with an aim to make clean India. The Mission aim is to provide upliftment to the maintenance of adequate disposal system of domestic sewage, cleanliness in rural and urban development, availability of safe and clean drinking water to all. Mission implementation and aim to complete sanitary India dream through Mission for the sanitary India by the year 2019. Some of the ministries involved in mission for the sanitary India are Ministry of Rural Development, Ministry of Urban Development, Ministry of Drinking water and Sanitation, Public Sector Undertaking, Corporations, National Government Organizations (NGOs), State Governments, Defence Research and Development Organisation etc. However, it has been clearly declared that the campaign is not only the duty of the government alone but each and every citizen of the country is equally responsible to keep the nation clean and green, and mission for the sanitary India.

Present paper discusses the progress of schemes documented under Mission for the sanitary India for the Aligarh city in Uttar Pradesh and problems encountered in the accomplishment of goal. This paper gives current scenario of India with respect to case study of municipal solid waste quantity, quality and its management.

Keywords: Mission for the sanitary India · Sanitation · Hygiene and sewage disposal

J. Calautit et al. (eds.), *Towards Sustainable Cities in Asia and the Middle East*, Sustainable Civil Infrastructures, DOI 10.1007/978-3-319-61645-2_5

1 Introduction

Our Prime Minister Narendra Damodar Das Modi launched an important project "Mission for sanitary India" on 2nd October 2014 on the birth anniversary of Mahatma Gandhi. This "Clean India Campaign" is the biggest ever cleanliness drives in the country. In which Prime Minister had appealed to each Indian citizen to get involved in the mission and make it successful. The main problem in our country is open defecation. More than 72% of rural people in our country are used to go open for excreting behind bushes, in fields or on roadsides. Which led to various other problems like untimely deaths of children due to spread of infections and diseases and most importantly rapes of young girls who go to deserted places to relieve themselves. We are more than 1.2 billion, out of which nearly 600 million people or 55% has no access to toilets. Even in areas, where there are toilets in rural India 32% of rural households, as per the 2012 data), facing the problem of unavailability of running water facilities. Slum dwellers in cities have no access to running water supply and toilets also. Considering this grave problem of rural sanitation in India and open defecation, the UPA Government had launched Nirmal Bharat Abhiyan in 1999. In this Abhiyan, a target was set for universal household sanitation coverage by 2012. This was an integral part of the Total Sanitation Campaign (TSC) launched in 1991. However, the Nirmal Bharat Abhiyan could not achieve its objectives. Nirmal Bharat Abhiyan has now been restructured by the "Mission for the Sanitary India" with the objective of making India a "clean" India by stopping the problems of open defecation, building up of toilets for all households, providing running water supply, treating of solid and liquid wastes in a proper manner including cleanliness of roads, pavements, and clearing of encroachments in unauthorised areas. Above all, the project aims at creating awareness among people about the need for proper sanitation and hygienic facilities.

Mission for the Sanitary India, with the help of the Government of India, political parties, NGOs, corporations and with active people's participation, is scheduled to be completed in 2019 which will mark the 150th birth anniversary of Mahatma Gandhi who laid great emphasis on cleanliness. He himself said "Sanitation is more important than Independence". It's been 67 years of Independence, and even today, more than half of India's population does not have proper toilets.

Municipal Solid Waste (MSW) is one of the major areas of concern all over the world. In developing countries like India, there is a rapid increase in MSW due to uncontrolled urbanization, rise in living standard, population growth, and industrial growth; (Joseph 2002). Expected generation of MSW until 2025 in India is 700 g per capita per day (World Bank Report-2006). The urban population of India is expected to grow 45% of total from the prevailing 28%. Hence the magnitude of MSW management (MSWM) problem is likely to grow to even larger proportions. The typical rate of increase of MSW generation in India cities is estimated around 1.3 annually (Bhide and Shekdar 1998; Shekdar 1999; Imura et al. 2005). Imura et al., observed that is developed countries generation like India of solid waste is more than developing countries. Imura et al., reported that MSW generation in less developed cities is 0.3–0.7 kg/capita-day. Rapidly developed cities is 0.5–1.5 kg/capita-day while for developed cities it is greater than 1 kg/capita-day in Asian countries like India. The quality and quantity of MSW generation by particular communities varies according to their social-economic status, cultural habits, urban structures, population and commercial activities etc. (Kansal et al. 1998; Singh and Singh 1998; Gupta et al. 1998; Rathi 2006; Jha et al. 2003; Ray et al. 2005; Sharholy et al. 2008). Planning, designing operation of MSW management system can be done on the basis of composition and the quality of MSW generated shown in Table 1.

Table 1. Composition of Municipal Solid Waste (MSW) (Source: AMC Agarwal et al. 2012).

Ingredients of MSW	Year 2000 (%)	Year 2025 (%)
Organic waste	40	60
Plastic	4	6
Metal	1	4
Glass	2	3
Paper	5	15
Other (ash, sand, grit)	47	12

Government world over the making efforts to improve solid waste management in their respective countries. There are about 593 districts and about 5,000 towns in India. About 27.8% of India's total population of more than 1 billion (as per Census 2001) lives in urban areas. The projected urban population percentage is 33.4% by the year 2026. The quantum of waste generated in India towns and cities is increasing day by day on account of its increasing population and increased Gross Domestic Product (GDP). The annual quantity of solid waste generated in India cities has increased from six million tonnes in 1947 to 48 million tonnes in 1997 with an annual growth rate of about 4.25%, and it's expected to increase 300 million tonnes by 2,047 (CPCB 1998). The annual population growth rate of India 2.15% and GDP growth rate is 9.3% (RBI 2006).

For proper execution of the Mission for the Sanitary India, a 19-member expert team has been formed under the chairmanship of scientist Raghunath Anant Mashelkar. Mashelkar is the former Director General of the Council of Scientific and Industrial Research (CSIR). The entire team will be involved to suggest the best and the most advanced technology to provide sanitation and water facilities in various states using methods that are affordable, sustainable, and amendable. On 2nd October 2014, when our Prime minister launched the mission, he was accompanied by party officials, Bollywood actor Aamir Khan, thousands of government employees, school and college students. Prime Minister has been actively supported by his cabinet ministers. To make it a mass movement, he has also nominated nine celebrities including Priyanka Chopra, Shashi Tharoor, Sachin Tendulkar and Anil Ambani etc., take up the cleanliness challenge, who in turn should invite nine more people and in this way the chain should continue. They have already accepted the challenge and have appealed more people to get involved. Some states have also participated in this cleaning campaign and more plans and programs are in the process to make this a success.

The present work attempt to bring forward the direness of the situation and also seek to find out possible remedial measure. For analysing actual distribution as well as simulating sustainable development in popular country like India where civic amenities and infrastructure development is still a problem between the numerous reasons.

Objective of Mission: The campaign of Mission for the Sanitary India launched by the government of India is aimed to accomplish various goals and to fulfil the vision and mission of "Clean India" by 2nd of October 2019. It has been expected that the investment to cost would be over 62000 crores of Indian rupee (US$ of 9.7 billion). It has been declared by the government that this campaign is taken as "beyond politics" and "inspired by patriotism". Important objectives of this mission are-

1. To construct individual, cluster and community toilets.
2. To eradicate the system of open defecation in India, to convert the insanitary toilets into pour flush toilets.
3. To remove the system of manual scavenging, to make people aware for healthy sanitation practices by bringing behavioural changes in people.
4. To link people with the programmes of sanitation and public health in order.
5. To generate public awareness, to build up the urban local bodies strong in order to design, execute and operate all systems related to cleanliness.
6. To start the scientific processing, disposals reuse and recycling of the municipal solid waste, and wastewater reclamation and cleaning the rivers.
7. To provide required environment for the private sectors to get participated in the capital expenditure for all the operations and maintenance costs related to the clean campaign.
8. Create public awareness about the drawbacks of open defecation and promotion of latrine use.
9. Lay water pipelines in all villages, ensuring water supply to all households by 2019.
10. Enhance improve the existing system to meet the clean water demand and in future expansion flexibility.

Implementation of Mission: The "Mission for the sanitary India" has two sub-Missions are as follows:

(1) **Mission for the Sanitary India (Rural areas):** For execution of these two sub-missions, the ministries of drinking water and sanitation and rural development will look after the affairs in rural areas and the ministry of urban development in urban areas. Under the Mission for The Sanitary India for the rural areas, the ministry of rural development will provide â,¹20 lakhs to each village each year for the next five years. Under this programme, the government has fixed the unit cost of individual household latrines at â,¹12,000 so that ample water supply can be provided for cleaning, bathing and washing. An estimated cost of â,¹1,34,000 crores will be spent by the Ministry of Drinking Water and Sanitation for the programme.

(2) **Mission for the Sanitary India (Urban areas):** The aim is to provide individual household toilets, community and public toilets, including solid waste management, covering â,¹1.04 crores households in all 4,041 statutory towns. It will

provide more than two lakhs seats of public toilets, and more than two lakhs seats of community toilets, including solid waste management facility for all towns. In areas, where there are problems of constructing individual household toilets, community toilets will be constructed. Public toilet facilities will also be made available in common locations such as markets, bus stations, near railway stations, tourist spots, and public recreation places. The urban development ministry has allocated â,¹62,000 crores for this project. The overall project cost has been estimated at â,¹1,96,009 crores. This amount will help in construction of â,¹12 crores toilets across the country. The ministries of rural and urban development have requested religious gurus and groups like Sri Sri Ravi Shankar, and the Gayatri Parivar to propagate the Mission for the Sanitary India project.

Major Issues of Mission

(a) According to Central Pollution Control Board (CPCB), Urban India generates about 47 million tonnes of solid waste every year. They reported that there is more than 75% of sewage or wastewater disposal which is neither treated nor recycled in India. Recycling of solid waste is a biggest problem. These issues need to be addressed now, so that the major crisis can be prevented in the future.
(b) In rural India, lack of adequate sanitation is a vast challenge.
(c) Another major challenge is to change the mind sets of the people to make their surrounding clean and hygienic.

The focus of the study area is on impact of solid and liquid waste due to non engineering and non scientific disposal. It is found that with increase in the global population and the rising demand for food and other essentials, there has been rise in the amount of waste being generated daily by each household. Waste is not properly managed in Aligarh city. Especially excreta and other liquid and solid waste from households and the community, are a serious health hazard and lead to the spread of infection diseases.

A Case Study of Aligarh City: The Aligarh city is famous for its Maratha fort, central university, "Aligarh Muslim University" and Lock manufacturing units. Aligarh is a medium size and densely populated city of North India, which lies in the shadow of country's capital New Delhi on 27 °53'N latitude and 78 °4'E longitude is presented in Fig. 1. Aligarh is the first major city situated in the western part of state of Uttar Pradesh on Delhi-Kolkata railway link and historical grand Trunk road. It is situated at a distance of 130 km, south east of Delhi the capital city of India in between the alluvial plains of holy river Ganga and Yamuna. The Aligarh district has an area of about 5,019 km^2 with the city occupying an area of about 34 km^2.

A Haryana based private agency A2Z had got the contract from of Aligarh Municipal Corporation (AMC) on a BOOT basis for period of 30 years. The company has been awarded contracts to handle 220 Tonnes per day (TPD) of MSW. The total project cost of â,¹34.17 crores is to be partly funded by debt, partly by equity and balance by way of capital subsidy from the Municipal Corporations. The company had tied up for a debt of â,¹11.50 crores and operations were effected from February 2011.

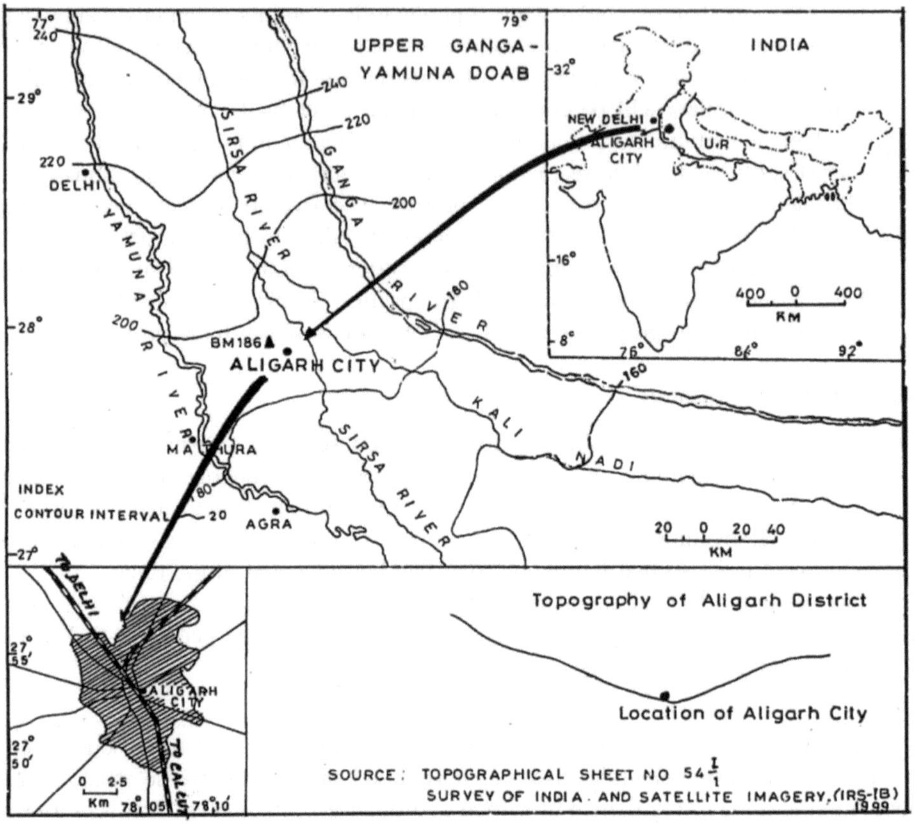

Fig. 1. Location map of Aligarh city, Uttar Pradesh, India. (Source: Survey of India and satellite imagery-1999)

The scope and operations of private company include Door to Door Collection (D2D), Transportation, Resource Recovery and Disposal of Remnants in Landfill. Private company came into existence in Aligarh in May 2010 and D2D collection started in June 2011.

However this firm did not achieved its aim and in January 2014, due to financial problem this firm had limited their work of SWM from 7 zones to only 2 zones. Remaining 5 zone of the city is covered by Aligarh Municipal Corporation (AMC) for its MSWM from January 2014. Currently AMC & A2Z both are working together for SWM and they are only collecting the wastes from the city.

Presently Aligarh has no proper disposal method. The wastes are being openly dumped onto low lying areas in the outskirts of the city. The disposal sites of Aligarh city were located within 5 km from the city limits. Some of the disposal sites are Chilkora, Quarsi, Mathura and Goolar road. The Disposal of a solid waste in a landfill induced burning the waste and this remains a common practice in most countries. Landfills were often established in abundant or unused quarries, mining voids or

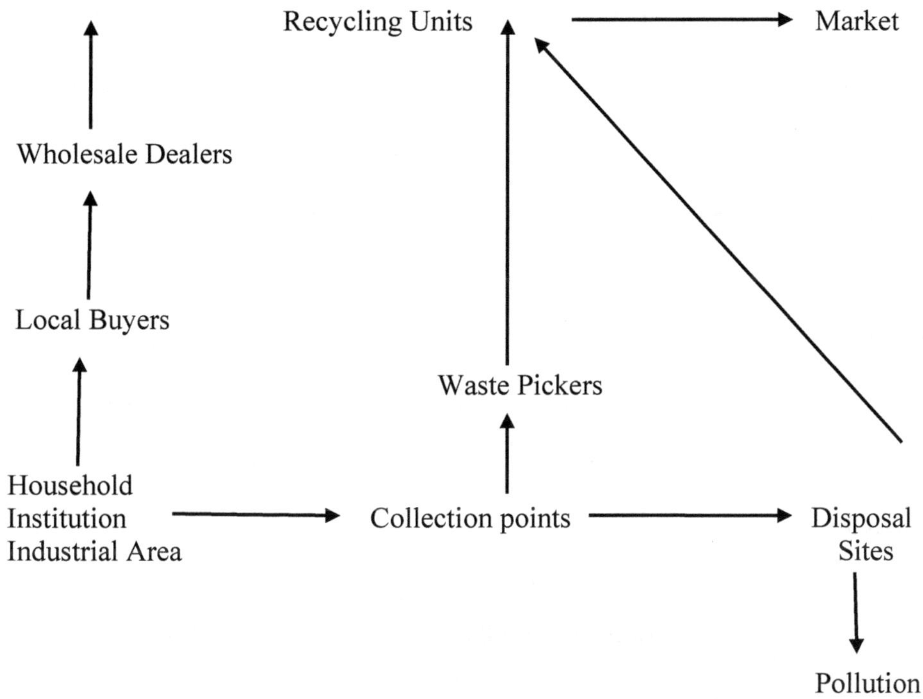

Fig. 2. Solid Waste Management (MSW) flow diagram in Aligarh city.

borrow pits. A properly designed and well managed landfill can be a hygienic and relating inexpensive method of disposing of waste materials.

The average quantity of MSW generated per day in Aligarh city like domestic, commercial, institutional, industrial, street waste quantity is presented in Table 2. This was measured by private agency under an assignment. The break-up in quantity in tonnes/day contribution to MSW quantity is given in Fig. 3.

Topography: Topographically the district of Aligarh represents a shallow trough, saucepan shape like appearance with the holy river Ganga in the northeast and river Yamuna in the northwest forming the highland peripheries. Aligarh city lies in this centre of the low-lying tract in Fig. 1, because of lack of natural drainage, rainwater and wastewater collects in the low-lying areas of the city. There is no outlet of water. The Aligarh city not only gets submerged during the rainy season but also some parts remain submerged all the year round because of imperfect natural drainage. The Aligarh city topography is saucer shaped that it lies in low laying areas having a Mean Sea Level (MSL) of 184.73 m and is surrounded by relatively high lying areas. The population of Aligarh city is about 1,33,729,723 based on the latest united nation estimates (3 January-2017). The population density is the district is 514 per km^2 and in the city 1, 4115 per km^2. Total number of slums in Aligarh city numbers 42,682 in which population of 258,841 resides. This is around 29.60% of total population of Aligarh city.

Table 2. Avarage quantity of MSW generated per day in Aligarh city. (Source: AMC Aligarh 2014)

Types of waste	Quantity in tonnes/day	Item	Percent by weight	Parameters	In % except pH and Calorific value
Domestic	280	Organic content	55.20	pH	6.4–8.3
Commercial waste	45	Inert/stone /ashes	19.60	Moisture content	15.0–25.0
Institutional waste	65	Paper	12.8	Volatile matter	28.0–30.0
Industrial waste	10	Plastic	4.5	Ash	40.0–45.0
Street waste	15	Wood	0.6	Fixed carbon	10.0–12.0
		Bones	2.3	Calorific value	1500–1800 (Kcal/Kg)
Total waste = 415 tonnes/day		Metal	1.8	Compostable matter	13.0–15.0

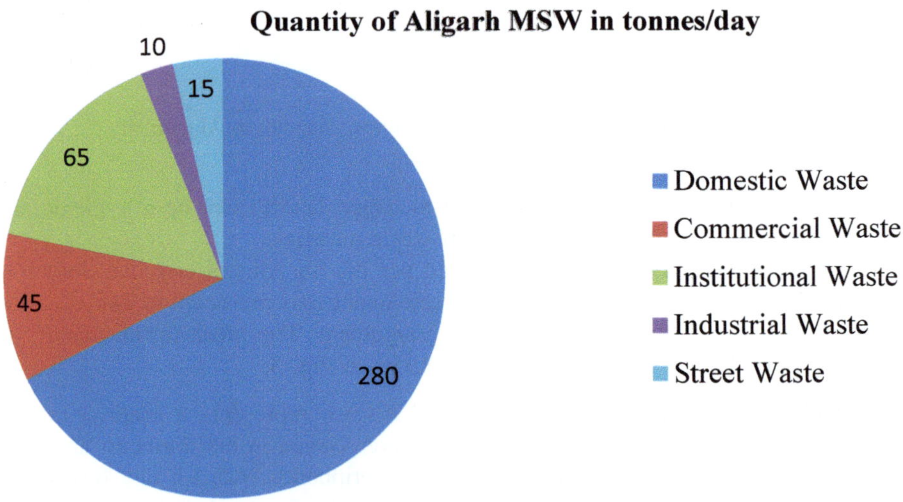

Fig. 3. Quantity of Aligarh MSW in quality generated per day in Aligarh city.

Climate and Rainfall: The climate is hot and dry in summer and cool and dry in winter with a rainy season in between the temperature in summer ranges from 37 °C to 4 °C and in winter from 10 °C to 1 °C. The city receives average annual rainfall of about 590 mm.

Problems in Aligarh City: Entire city is suffering from actuate household garbage and rubbish, street sweeping, construction and demolition debris, sanitation residues, trade and non hazardous industrial refuse and treated biomedical solid waste, open defecation near railway line and colonies is scattered all around the city due to extremely poor waste management is shown in Figs. 4, 5 and 6 These waste scattered in city can be seen in Figs. 8 and 9. The Aligarh city has been generated basically three types of waste like solid, liquid and medical waste. The solid waste see all round scattered on road, lane, near school, and civic offices, liquid waste all most all home drains are open outside, there is no proper facility and medical waste generated

Fig. 4. Open defecation in Auto stand near Tasveer Mahal, Aligarh Muslim University road, Aligarh

Fig. 5. Open defecation in Railway line, Aligarh-Junction, Aligarh.

Fig. 6. Person not use the toilet shows lack awareness in near Aligarh Junction.

hospitals, clinic and other medical facilities. The study revealed that there is an active informal sector collecting recyclable materials at various stages from individual households to final disposal sites is shown in Figs. 10 and 11. The majority of the materials that can be recycled or reused are recovered from the individual households by itinerant buyers. Waste that finds way onto the road side open dumps or bins is picked by waste pickers. Generally women and children are involved in waste picking activities. The buyers and waste pickers sell their recyclable materials to the middlemen who then sell them to the different recycling units. A typical solid waste management materials flow diagram for Aligarh city is shown in Fig. 2. The Aligarh MSW waste consist of Organic content (55.2%), inert material (19.6%), paper (12.8%), plastic

Fig. 7. Both child's open defecation near slum, Shahjamalpur, Aligarh.

Fig. 8. Unwanted and discarded materials near Maris Road, Aligarh

Fig. 9. Solid waste and discarded materials in drain affect the Ecosystem, near Jamalpur, Aligarh.

(4.5%), bones (2.3%), textile (2.7%), metal (1.8%) and wood (0.6%). Estimate 30% of the total waste is being recycled. Approximately 80% of the collected waste is disposed as open dumping. There is no sanitary landfill in Aligarh city.

Methodology: A medium size city, Aligarh, having population of about 0.8 million and situated 130 km from the capital city of India, Delhi was selected for this study. A theoretical framework and a protocol; were developed to conduct this study. Essential information needed within the framework and protocols were collected from various sources. The study focus on Mission for the Sanitary India is completely ineffective or not in Aligarh city Uttar Pradesh, India. Figures 8 and 9 shows solid and liquid waste of different location in study areas such as Jamalpur, Shahjamalpur,

Fig. 10. Accumulation of solid waste and waste water in and around residential areas, creating ideal breeding ground for insects and pests near Pratibha Colony in front of Nagar Nigam Workshop, Aligarh. (Dirty still water)

Fig. 11. Liquid waste generated from factories and other sources in front of Govt. office, near Naurangi lal School, Aligarh

Jeevangarh, Tasveer Mahal, Dodhpur, Naurangabad, Maris road etc., which face a lot of problems related to solid and liquid waste have been collected, and secondary data were collected directly from concerned office, research papers, newspapers, journals and articles etc. Personally make visit in the city and to the waste disposal site to assess the solid waste management system.

Aim of this work was to analyze the present status and future challenges of solid waste management system in India. Within this aim Aligarh city was examined for its solid waste management system.

2 Result and Discussion

Study has revealed that the amount of municipal solid waste collected by the private body in Aligarh city is one of the order of 415 tonnes per day. Present practices of SWM in Aligarh city are very weak. Approximately 80% of the collected wastes are disposed as open dumping. There is no sanitary landfill in Aligarh city. The city is no exception and management of solid waste disposal is passing a problem of this rapidly developing city of western Uttar Pradesh, on most accounts. Aligarh city is regarded to be one of the most polluted cities of the country due to its cultural and physical setting. Primarily, an old city like Aligarh with low relief and slope, with dense urbanization and narrow pathways and ever increasing uncontrolled population load, modern civil engineering projects with state of art GIS and GPS technology can be prove to be a boon for planning, maintaining, formulating, and simulation along with models that may come out for next 10, 20 and 50 years. Sustainability requires most modern technology tool like LIDAR survey along with total station survey of entire city and thus a chain of all sanitary like can be made which can be further joined and rejoined keeping with the slope aspect minimal. A GIS database generated inform of polygon, point and line with feature information of number of household, family groups linked up can serve to plan for other civil projects such as all underground sewage system or installation of real time monitoring system at selected places.

Rainfall in Aligarh occur in between June to September where additional rainwater gets drained into sewage system with no scope of conserving them, thus minimize the chances of ground water recharge that cannot percolate through concrete urban state and flows forward or gets evaporated. For sustainability sanitary system of India should be aimed at rain water harvesting as well, for old cities or urban centres like Aligarh. Civic projects structure design and methodology for sanitation and efficient waste treatment with techniques and methods of rainwater harvesting for recharge of aquifers accordingly to geological standards for sustainability adopted for Aligarh can be adopted all over the globes with similar condition.

Environmental sanitation envisages promotion of health of the community by providing clean environment and breaking the cycle of diseases. It depends on various factors that include hygiene status of the people, types of resources available, innovative and appropriate technologies according to the requirement of the community, socio-economic development of the country, cultural factors related to environmental sanitation, political commitment, capacity building of the concerned sectors, social factors including behavioural pattern of the community, legislative measures adopted, and others. India is still lagging far behind many countries in the field of environmental sanitation.

Health and disease related issues in Aligarh slum area Fig. 7 adversely affect the health status of inhabitants due to lack of basic infrastructure and health services. Overcrowding in slums is common cause of psychological stress and increases the rate of disease transmission due to frequent contact. There is intense need for addressing vector born, water born diseases, in informal settlements and mobilization of health services for these urban poor. The health and medical facilities in the households of slum was found to be negligible. The accessibility and utilization of the healthcare services among a population from community was very poor. The illnesses among children aged under-5 years in study region were identified as respiratory tract infections, diarrhoea, malaria, dengue, skin problems and malnutrition.

The Aligarh city people is not aware of the poor waste management which will lead to various diseases diarrhoeal diseases, cholera, typhoid fever, amoebiasis, hookworm infections, conjunctivitis, and malaria, acute upper and lower respiratory infections, Chronic obstructive pulmonary diseases, asthma, prenatal mortality, pulmonary tuberculosis, low birth rate weight, eye irritation and contract, AIDS, Hepatitis B and C among the households during the last 2 years, (J.N. Medical collage Report 2014). The entire Aligarh city is suffering due to large amount of solid waste management problems because non existence of nongovernmental organization, non availability of well qualified persons among the Aligarh, lack of education, lack of public awareness etc. but we can improve or resolve the poor waste management issues from Aligarh city if development of National Government Organisation (NGO), development of charity, general awareness of waste and its impact of human life.

We can improve if we take the responsibility and work together to the Aligarh city if we development of NGO/charity based on the donation, NGO can work along Aligarh Municipal Cooperation, general awareness of water and its impact on human life etc.

3 Conclusion

The MSW generated in Aligarh city depends upon the population, climate, uncontrolled urbanization, socio-economic condition. The entire Aligarh city is suffering due to acute waste management problem, even though it is not being addressed by the single community person due to non existence of NGO, Aligarh community is careless towards the responsibilities to clean city, lack of general awareness regarding effect of waste is very poor among the residential. Aligarh MSW has a very similar composition to that of other medium-sized Indian cities. Organic content in the waste was found highest organic content (55.20%), Inert (19.60%), paper (12.8%), plastic (4.5%), cloths (2.8%), bones (2.3%), metals (1.8%), wood (0.6%), and others (0.4%), shown in Table 2. The study has revealed that collection efficiency of the private operator and Aligarh Municipal Corporation is approximate 80%. According overall aspects shows that the solid waste management in Aligarh city has been significantly improved after the involvement of private sector.

A Nationwide Real Time Monitoring system has also been launched by the Government of India in order to construct toilets under this campaign aiming to attain 100% open defecation free India by 2019. We should participate in this mission by meeting our hands together and taking this cleanliness mission as an important part of our life with the vision of clean India achievement through the effort of each and every Indian citizen and not by the effort of government only. Beside all these the government imposed Mission for the Sanitary India Cess at the rate of 0.5% on all services, which are presently liable to service tax from 7th November 2015. This will also be helpful in achieving the goal of cleanliness and hygiene by the due date 2nd October

2019 on 150th anniversary of our beloved Father of Nation Mahatma Gandhi and towards his biggest dream.

"Clean India Green India: Dream of Mahatma Ghandi"

The view and comments expressed and notified in the paper are the personal view of the author and suggestion and critical review will be appreciated.

Acknowledgement. First of all I would like to thanks Honourable Prime minister Narendra Modi to continue the dream of Mahatma Gandhi Make in India mission "Mission for sanitary India". The authors acknowledge all the persons involved in Aligarh Municipal Corporation (AMC) for providing all the pertinent information. The authors would like to acknowledge to thanks of staff of Mangalayatan University specially Dr. (Mrs.) Monika Singh, Mr. Mohammad Baquir, Mr. Waseem Ahamad Khan and Dr. (Miss) Shamim Ajaz, Research Fellow, Jamia Hamdard University, New Delhi. I am also thankful to my both students Shahbaz Shamshad Khan and Divyanshu Vats which are the Civil Engineering students and doing minor project of "Mission for the Sanitary India" under my guidance.

References

Agarwal, D., Gupta, A.K., Agarwal, M.K.: Analysis of solid waste management strategies in major cities of Uttar Pradesh in India. J. Pure Appl. Sci. Technol. **2**(1), 66–74 (2012)

Bhide, A.D., Shekdar, A.V.: Solid waste management in Indian urban centres. Int. Solid Waste Assoc. Times (ISWA) **1**(1), 26–28 (1998)

Central Pollution Control Board of India (CPCB): Status of solid waste management in metro cities. CPCB (1998)

Gupta, S., Krishna, M., Prasad, R.K., Gupta, S., Kansal, A.: Solid waste management in India: option and opportunities. Resour. Conserv. Recycl. **24**, 137–154 (1998)

Imura, H.Y., Shinirakawa, H., Memom, M.A.: Urban environmental issues and trends in Asia- an overview. Int. Rev. Environ. Strat. **5**, 357–382 (2005)

Jawaharlal Nehru Medical Collage (JNMC), Health report (2014)

Jha, M.K., Sondhi, O.A.K., Pansare, M.: Indian J. Environ. Prot. **23**(10), 1153–1160 (2003)

Joseph, K.: Perspectives of solid waste management in India. In: International Symposium on the Technology and Management of the Treatment and Reuse of the Municipal Solid Waste (2002)

Kansal, A., Prasad, R.K., Gupta, S.: Delhi municipal solid waste and environment -an appraisal. Indian J. Environ. Prot. **18**(2), 123–128 (1998)

Rathi, S.: Alternate approaches for better municipal solid waste management in Mumbai, India. J. Waste Manag. **26**(10), 1192–1200 (2006)

Ray, M.R., Roychoudhury, S., Mukherjee, G., Roy, S., Lahiri, T.: Int. J. Hyg. Environ. Health **108**(4), 255–262 (2005)

Reserve Bank of India (RBI): Macroeconomic and Monetary Developments: First Quarter Review, Ministry of Finance, Government of India, New Delhi (2006)

Sharholy, M., Ahmad, K., Mahmood, G., Trivedi, R.C.: Municipal solid waste management in Indian cities - a review. Waste Manag. **28**(2), 459–467 (2008)

Shekdar, A.V.: Municipal solid waste management - the Indian perspective. J. Indian Assoc. Environ. Manag. **26**(2), 100–108 (1999)

Singh, S.K., Singh, R.S.: A study on municipal solid waste and its management practices in Dhanbad-Jharia Coalfield. Indian J. Environ. Prot. **18**(11), 850–852 (1998)

World Bank Report: Practical Guidebook on strategic Planning in Municipal Waste Management (2006)

The Free State Public Transportation System: A Comparison Between Buses and Taxi Services and Adaptation of ICT Solutions

N. Bashingi[1(✉)], M. Mostafa Hassan[2], and E.M. Masinde[3]

[1] Sustainable Urban Roads and Transportation Research Group, Department of Civil Engineering, Central University of Technology, Bloemfontein, Free State, South Africa
nbashingi@gmail.com

[2] Department of Civil Engineering, Central University of Technology, Bloemfontein, Free State, South Africa
mmostafa@cut.ac.za

[3] Information Technology Department, Central University of Technology, Bloemfontein, Free State, South Africa
emasinde@cut.ac.za

Abstract. This paper explores Information Communication Technologies (ICT) for consideration in the public transportation system in the Free State and how the two main types of public transportation; buses and taxis, are likely to adapt to these technologies. Taxis and buses operation and management of operations are completely different; their differences are likely to affect how each adapts to these technologies. Adaptation to the integrated public transportation system by the bus and taxi users, drivers, and operators will be key to successful ICT integration. Focusing on the benefits that ICT solutions have brought in some regions in the world should not overshadow consideration for possible slow adaptation, lack of acceptance or resistance by stakeholders when applied in a different environment and a different public transportation system.

This paper considers the different operational scenarios, government involvement and political will, public – driver – operator participation and willingness and ICT awareness. Whether the technologies and applications will be suitable and adaptable to the existing system and their perceived influence on adaptation to an ICT integrated public transportation system from implementation until there are fully embedded into the system are also considered. Feasibility of some ICT solutions, to the decision makers and operators, considering financial, technical and management feasibility and degree of integration will influence the adaptation process.

Keywords: ICT · Public transportation · Adaptation

1 Introduction

ICT solutions have been successfully implemented in public transportation systems all over the world and it is evident that they have led to improved quality of public transportation services in both developed and developing countries such as Kenya,

J. Calautit et al. (eds.), *Towards Sustainable Cities in Asia and the Middle East*,
Sustainable Civil Infrastructures, DOI 10.1007/978-3-319-61645-2_6

United Kingdom, Netherlands, Singapore and in South Africa, Cape Town (Ministry of Transport, Public Works and Water Management 2010). In these places, ICTs are used for different modes of public transportation, i.e. mini-bus taxis, buses and trains. These integrations started over the years with the oldest technologies such as telephones. Most countries and cities' public transportation systems have since developed and adapted to new technologies, some of which they abandoned along the way, kept using and upgraded or adopted newer technologies. An integrated public transportation system, in this context, refers to bus and taxi services using Information Communication Technologies, which may include, but not limited to the following, provision of real time information, easy online and mobile payment methods, trip planning, Automatic Vehicle Location technologies. These technologies have been made available to stakeholders and the systems have adapted to the changes in some cities. Even though African countries and cities like Nairobi and Cape Town have managed to successfully integrate certain technologies into their public transportation, Idongesit and Knud (2014) highlight the concern and fear that African countries cannot possibly catch up with the ICT development and deployment that already prevails in other continents as they cannot use the same approaches to solve the challenges they are facing.

Increased development and use of ICT such as electronic fare payments, real-time information provision through websites, mobile applications, social networking platforms in public transportation systems exposes the non-existence of ICTs in the Free State public transportation system. The province will have to catch up with the rest of the world by adopting these trends and technological advancements in order to cater for increasing demands in public transportation; information, reliability and convenience. The current, conventional public transportation services; buses and taxis will therefore have to adopt and adapt to the use of ICTs if ever integration takes place.

1.1 Study Area and Public Transportation Limitations

The study was carried out in the Free State province, South Africa; the third largest province covering 10.6% of the country's surface area. The Free State province has a population of 2 745 590 (Statistics South Africa 2012), with 2.6 million people living in the urban areas in the province and about 0.6 million living in the rural areas (Toba et al. 2012).

Public transportation in the Free State province is faced with a lot of challenges: the low number of people who use public transportation among others. There is still a large number of people reluctant to use public transportation, 30.2% workers in the province use public transportation (buses and taxis) while 34.9% use private transportation. (Statistics South Africa 2014). More workers who travel on average five days a week to work use their own vehicles. The only modes of public transportation considered for this study are buses and mini-bus taxis as these are the main public transportation service providers.

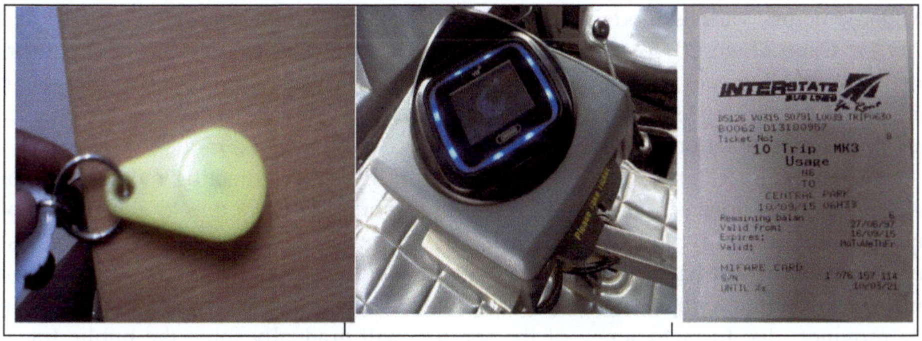

Fig. 1. Ticketing system in interstate buses

1.2 The Differences Between Bus and Taxi Services in the Free State

Currently, the only public transportation bus service provider providing regular or conventional transportation services in the Free State is Interstate Bus Lines (IBL), which is a private company subsidised by the government through the Department of Transport. IBL operates in the Mangaung Metropolitan Municipality's major cities, Bloemfontein, Thaba-Nchu and Botshabelo. ICT solutions that are part of the company's services include a payment system in which passengers pay to load money into their tags, which they tap on a machine when entering the bus to get their ticket (Fig. 1). Travel information, i.e. timetables on PDF format are provided for downloading on the company's website and any information to be communicated by the company to passengers is conveyed through an intercom, this only takes place at the main bus-terminal in Central Park.

Taxi services are privately owned and operated by individual taxi owners and small companies. Vehicles used as taxis are mini-bus taxis normally sixteen seaters. Technologies often found in these taxis are tracking devices and entertainment devices such as a radio or Mp3 players. These do not mostly benefit the passengers, i.e. because they do not provide them with any information, and are mostly for the driver's entertainment. Taxi fare payments are made in cash before or during travel and there is no real-time information available. Taxis cover remote areas within the province and are cheaper compared to buses, the limitations in bus services include the fact that it doesn't cover smaller towns and villages in the province and for people travelling to and from those areas, their only available mode of public transportation is mini bus taxis.

2 Literature Review

Some cities' public transportation systems have adopted ICT solutions for different types of vehicles; trains, mini-bus taxis and buses. In Nairobi, Kenya for instance, they use mini bus taxis known as Matatu and fares are paid using a cell phone through M-Pesa services (Mulipi 2015). In London, Perth and Cape Town, ICT solutions are available across various modes (Transport for London 2012; NNT DATA 2015 and City Of Cape Town 2015), which consist of electronic fare payment systems, information display screens, online booking and payment systems. Success of the above cities and countries systems is due to stakeholder acceptance which led to quicker adoption and adaptation.

Adaptation to ICT may affect travel behaviour by either attracting more people or change the behaviour of the existing public transportation users due to availability of information, reliability and convenience. The efficiency of public transportation systems over the world have improved the systems and attracted more users. Growth in public transportation improvement globally and a decline in public transportation use in the province, due to inefficiency of the service, will gradually force the public transportation system to change. Adoption of ICTs and adaptation to an ICT integrated public transportation system, for buses and taxis will be determined by the following factors:

2.1 Political Will, Legislation and Policies

The role of politics and government in public transportation policy making and operations affects the development of the system. Public transportation operations globally are influenced by politics; the Free State province therefore is not an exception. South African legislation on public transportation decision making gives political leaders the power to control and make important decisions regarding the public transportation system, that is, ministers and MECs (Member of Executive Committee) (Department of Transport 2000). It is the duty of the minister to prescribe requirements for integrated fare systems, comprising fare structures, levels and technology, to ensure compatibility between such systems. This highlights the issue of political will; implementation happens because the people in power to pass authority and orders for prescriptions to be implemented are willing to do so, therefore until the minister permits or deems necessary, integration of fare systems may not happen.

Even so, policies have been made before and the parties involved did not implement these policies entirely. Government legislation and policies such as the Taxi Recapitalization Programme and provincial (or cities) integrated transportation plans most often include ICT components which are seldom implemented. The Taxi recapitalization programme (TRP) introduced in 2001 in which old taxis were scrapped off the roads in exchange for new ones was supposed to include an Electronic Management System (EMS). This system was to include Electronic Fare Collection System using smart cards, Monitoring and Control System, and Vehicle Identification Tracking and Recovery System. This programme however, has mostly resulted in replacement of vehicles with only some or none of the EMS components.

TRP would have speeded the adoption of ICT by taxis, but it was and still is not fully implemented, stressing the fragmentation and lack of coordination in the taxi industry, together with the lack of cohesion between government and taxi owners or industry.

2.2 Stakeholder Acceptance

The public transportation system involves many stakeholders: the operator (the bus company or taxi owner and the associations they belong to), the driver, the passenger, the government and the general public. Acceptance of the technologies by all stakeholders may result in the system adapting to the innovations faster. The technology acceptance Model by Davis, Bagozzi and Warshaw (1989) articulates perceived usefulness and ease of use of technologies as the main motivators of acceptance of technologies. If technologies adopted for use in the Free State fit these driving forces, adaptation will be easier. There are also debates regarding acceptance of technologies, that users accepting technologies does not necessarily mean that they will support them (Regan et al. 2014). Further, lack of knowledge about technological developments, how they function and acceptance of the proposed systems and their possible impacts after implementation may also affect adoption (Geenhuizen and Thissen 2002). Stakeholder's attitudes and perceptions towards technologies can also impact the adaptation process especially from the beginning, but Van Exel and Rietveld (2010) suggest that perceptions can be changed and this may lead to changes in attitudes.

2.3 Finances

Deployment of ICT in itself requires financial obligations in order to be successful. The Interstate Bus Lines company, which provides all public bus services to passengers in the Free State province is a private company, which is also subsidized by the government. On the other hand, taxis are operated as small businesses by individuals or small companies. The taxi operators finance themselves, therefore for them to integrate any technologies on their own, they will have to do that from their own financial resources. However, in the case of TRP, EMS were supposed to be compulsory in all taxis as the taxis would have been designed specifically for the programme; hence, replacing the old taxis with shared financing by the government and part of the fee from the taxi owner. The research indicates that most of the current taxis being used only have (VITRS). Fares are still payed in the form of cash; this indicates that the programme which was to be completed in 1999 have not been able to achieve its intended purposes. This implies that if ICTs were to be installed now, this has to be at the expense of the taxi owner, be it individuals or companies. This puts bus service provider in a more favourable financial condition to be in a position to integrate ICT solutions into their operations than taxis. The Taxi Recapitalisation Programme also proposed that after the recapitalisation process had paid off, taxis should be in a suitable position to apply for government subsidies.

2.4 Operating Environment

IBL Bus service operate in Mangaung Metropolitan Municipality, the only metropolitan municipality in the province, taxi services operate in the cities and the rural areas within the province. The population density in areas within the province would also influence schemes such as ICT based Demand Responsive Transportation schemes. The population in rural areas which has lower literacy rate than the urban areas is likely to adapt slowly to the new system. Buses and taxis operate under entirely different conditions.

3 Research Method

Both primary and secondary data collection methods were used to quantitatively and qualitatively gather data using questionnaires, interviews and literature review.

Literature review was carried out in order to get the background of ICT and public transportation. Public transportation operators, drivers and passengers were interviewed. The questions were based on ICT awareness by the stakeholders, their likelihood to use public transportation after ICT solutions are integrated into the public transportation system and technologies the operators are willing to incorporate into their services. Regression analysis was used to mathematically correlate demographic, current public transportation and ICT related variables with likelihood of people using public transportation after ICT integration, which resulted in a model forecasting future public transportation usage. Determining whether people will accept ICT into the public transportation system whereas the technologies used will affect the stakeholder's perceptions to these technologies, their acceptability and feasibility.

A sample of 162 respondents was used, amongst them 50 mini bus taxi operators and 1 representative of the bus company operating in the Free State province was used. The respondents include public transportation users and non-users, taxi owners, taxi drivers who were randomly picked. ICT experts, transportation planners and academics in both transportation and ICT were identified based on their expertise and interviewed.

3.1 Research Aim

This study evaluates the possible adoption of ICT solutions into the current public transportation system in the Free State province and compares the two main modes of public transportation's presumed future adaptation based on the stakeholders needs and the service providers' willingness to integrate ICT into their operations, under the premise that there will be differences in rates of adoption and adaptation for buses and taxis.

4 Results

4.1 ICT Awareness and Accessibility

Awareness and accessibility of ICT are presented in Table 1, which shows the reasons for lack of internet access by public transportation users. People's willingness to use internet if it was available gives hope that with more people having access to the internet, it will mean possible quick adaptation by users of both taxis and buses. The public's willingness to continue using public transportation after ICT integration is a possible sign of quick adaptation; this means that for public transportation users, the introduction of ICT solutions is a long overdue development.

Table 1. Internet accessibility in the free state

Reasons for lack of internet access	Percentage
Do not need it	2.8
Expensive	30.6
Not interested	16.7
Willing to use if available	50
Total	100

4.2 Future ICT Uptake

Technologies available in the some of the buses used for public transportation are illustrated in Fig. 1. These are the only ICTs used, being an electronic ticket machine and a tag. Figure 2 also shows technologies respondents indicated are available in vehicles, especially taxis. Majority of the devices are for entertainment purposes (music and DVD players), followed by cell phones which are the driver's personal phones which may be used for communication between the driver and owner. Vehicle destination display (mostly used in buses) and Global Positioning System (GPS) devices are available in some taxis. The types of devices available indicates that, even with the TRP, taxis did not install all ICT components listed under the EMS. Public transportation service providers are willing to provide technologies which benefit them, based on security reasons especially for their vehicles and taxi fare, not those of the public transportation users.

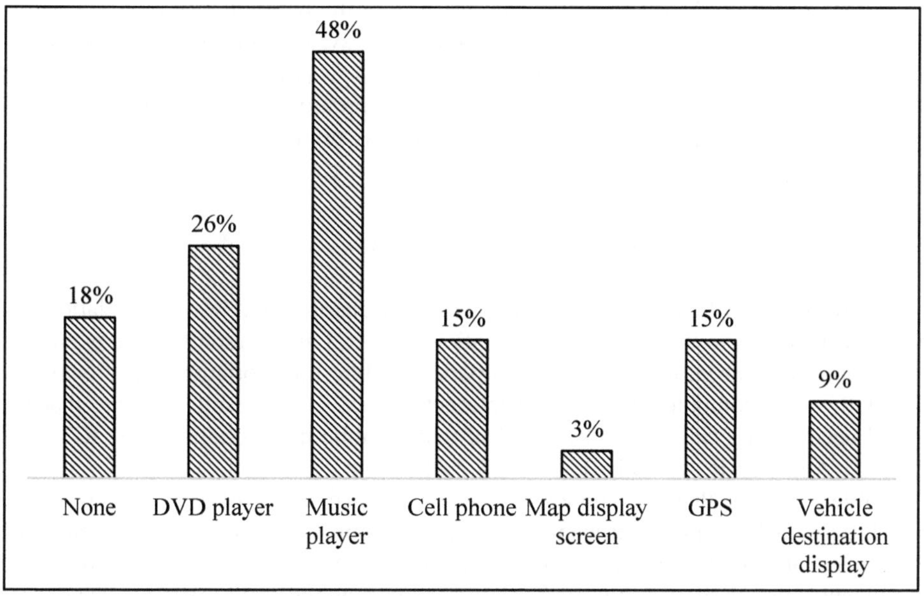

Fig. 2. Devices currently available in public transportation vehicles in free state

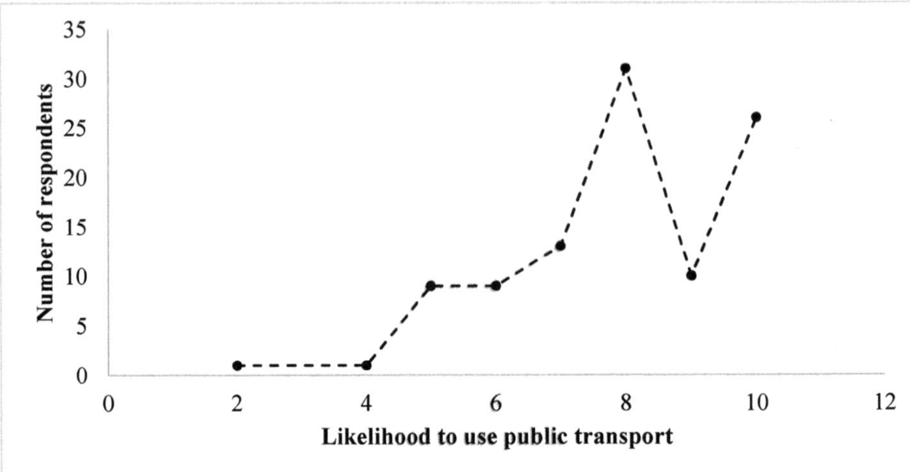

Fig. 3. Users likelihood to use public transportation after ICT integration in the free state province

There are possibilities of a vast majority of the population in Free State, resisting the new technologies; majority of people who use taxis in rural areas are not aware of ICT and are not computer literate. Lack of awareness, if individuals are not willing to learn or no measures taken to increase awareness may lead to high levels of unacceptability. Accessibility of internet will also play a large part in the adaptation process.

Figure 3 was used to predict the need for ICT in the public transportation system, the likelihood of respondents using the transportation system after ICT integration was measured, using a 10-point scale. A mean of 7.94% was recorded. This shows that there are high chances of people using public transportation after ICTs are introduced. Further analysis carried out using regression analysis produced the following results:

Regression model

$$Y = \alpha + \beta_1 x_1 + \beta_2 x_2 + \beta_3 x_3 + \beta_4 x_4 \ldots\ldots\ldots + \beta_p x_p$$

An R^2 value of 0.551 was produced, which means that 55.1% of the relationship between likeliness to use public transportation after ICT integration, can be explained by the independent variables; Age, Occupation, Public transportation usage, satisfaction with taxis, computer literacy and internet access. These results indicate that the model accounts for 55.1% of likelihood of using public transportation after ICT integration. (Adjusted R^2 value is 0.382 = 38.2%) (Table 2).

Table 3 shows that the model has a significance value of less than 0.05 i.e. $p < 0.05$; with R^2 being significantly greater than 0 proving that the age, occupation, PT usage, satisfaction with taxis, computer literacy and internet access account for a significant amount of variance in likeliness to use public transportation (dependent variable) and the overall regression model is significant: $F\ (6, 16) = 3.268$, $R^2 = 0.551$.

Table 2. Model summary output

Summary output	
Regression statistics	
Multiple R	0.742
R Square	0.551
Adjusted R square	0.382
Standard error	0.678
Observations	23

Table 3. ANOVA

ANOVA					
	df	SS	MS	F	Significance F
Regression	6	9.0020	1.5003	3.2685	0.0271
Residual	16	7.3445	0.4590		
Total	22	16.3465			

The final coefficients derived presented the following model:

$$y = \alpha + \beta_1 x_1 + \beta_2 x_2 + \beta_3 x_3 + \beta_4 x_4 + \ldots\ldots\ldots + \beta_p x_p$$

Therefore; Likelihood to use public transport after ICT integration = 5.201

$$-(0.185\ \text{Age}) - (0.314\ \text{Occupation}) + (0.350\ \text{Public Transportation Usage}) + (0575\ \text{Satisfaction with Taxis}) - (0.214\ \text{Computer Literacy}) - (0.145\ \text{Internet Access}).$$

These results when simulated show that people with higher satisfaction levels with the current taxi services, computer literate and have access to the internet are more likely to use the ICT integrated blic transportation system.

Based on the variables used and their relationship, which the dependent variable relies upon human behavior and some of the independent variables are based on individual's feelings towards the existing transportation system, the possible explanation to this are the different perceptions people might have towards adopting ICTs, which will impact how they adapt to the new ICT integrated public transportation system.

4.3 ICT Technologies

Technologies preferred by users and those the operators are willing to provide are shown in Figs. 4 and 5, the ones users are willing to use will be more acceptable for either buses or mini bus taxis. Long distance taxis, which normally leave only when full would use online booking compared to local taxis because people are always

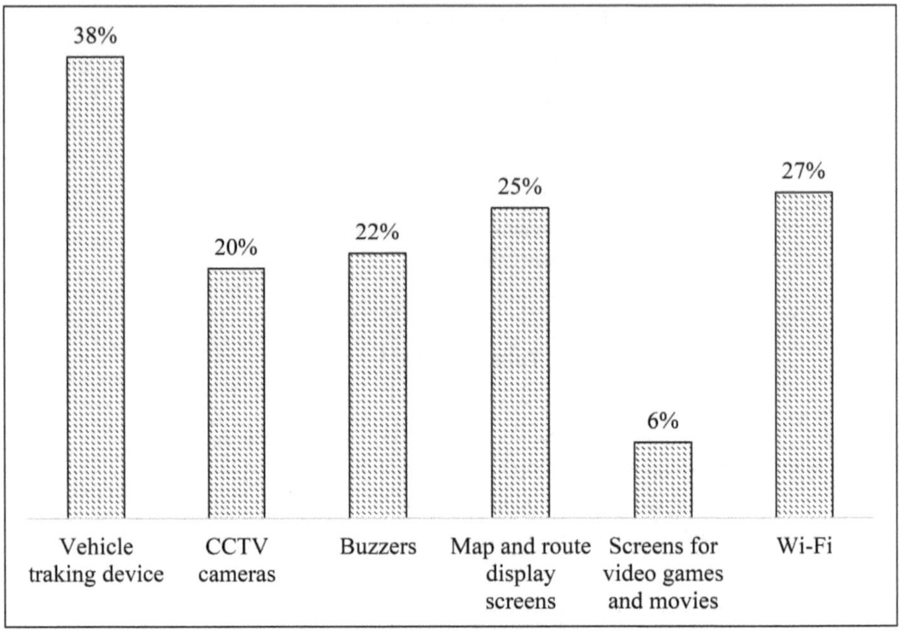

Fig. 4. Technologies respondents are willing to use in public transportation when made available

travelling locally, within cities which would not require prior online booking. Online booking of seats wouldn't also carry financial implications for the operators and passengers. Bus services, especially travelling to Botshabelo and Thaba Nchu at peak hours would also benefit from online booking.

Vehicle tracking devices, CCTV cameras and speed monitors, when deployed monitoring technologies would be acceptable to operators (vehicle owners). However, drivers, especially taxi drivers would consider them to be invasion of their privacy. This is mainly because taxi drivers unlike bus drivers, tend to use the vehicles for their own errands and tracking and monitoring them at all times would compromise their freedom. There are higher financial implications of tracking and monitoring vehicles as they use more advanced technologies. On the other hand, passengers for their safety would accept such technologies. The taxi system's safety is still questionable due to reasons such as taxi driver behaviour, hijacking of taxis and robberies and thefts at taxi ranks. Operators and passengers would be willing to accept these technologies while drivers are likely to reject them.

Countdown timers, available seat detectors; the size and capacity of buses allow for installation of seat detectors. Passengers before settling down, especially in large articulated buses walk down the bus looking for available seat; seat detectors will help passengers identify empty seats from the vehicle's entrance. On the contrary, mini bus taxis are small and a passenger can easily identify an empty seat from the entrance. Therefore, these technologies will be much more suitable and acceptable for buses.

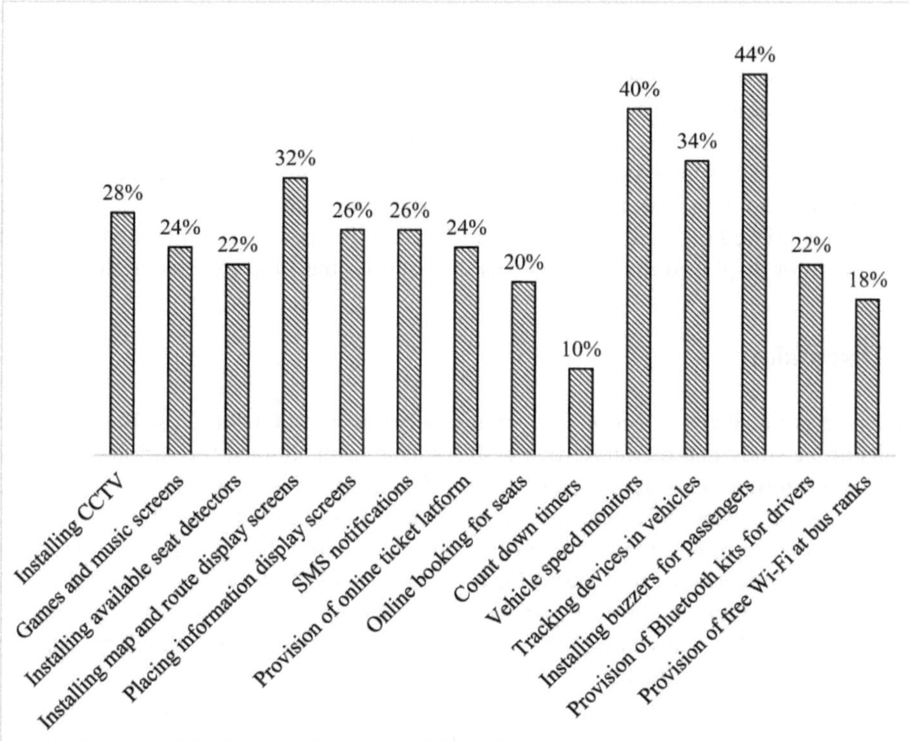

Fig. 5. Technologies public transportation service providers are likely to incorporate in their service in future

SMS notifications, for both bus and taxi services, the use of SMSes to notify waiting passengers of delays in schedule or vehicle breakdown will be both affordable to the operators and acceptable. Even though taxis are always available in their designated routes in towns, in cases of delays the next taxi would pick up the passengers, buses run on schedules and passengers might have to wait longer waiting for another bus to be dispatched.

Map and route display screens, entertainment screens, entertainment screens for video games or television would be acceptable to younger passengers and considered a nuisance or distraction by older passengers. The effectiveness of these will depend on the size of screens to be used; buses and mini bus taxis can accommodate screens internally. Screens are common in some taxis and buses for entertainment purposes. The same screens can be used for displaying real time map and route information to passengers. Using the existing devices would result in less financial implications to the vehicle owners or the state. How compatible the screens are with the applications or software used however may determine the overall financing and implementation. Requirement for new screens would mean more financial burden for the vehicle owners, companies or the state.

Provision of Wi-Fi at bus ranks and stops, as internet accessibility is currently a problem, providing accessible internet at bus stations will be an acceptable technology.

Destination display LED signs, already exist on some of the Interstate Bus Line buses, in the form of LED display boards in front of the vehicles. Passengers are already accustomed to this technology. Mini bus taxis on the other hand do not have their destinations displayed on them, for a passenger to know where the taxi is going, they have to stop it and ask the driver. However, some of the mini bus taxis have their destinations printed on them. LED signs are feasible in taxis; either mounted on top of the vehicles or displayed the same way as in buses, on the front exterior of the vehicle.

5 Discussion

Buses are more likely to adopt ICT solutions before the taxi services. Based on the results of the study and literature review, ICTs have been adopted for various modes of public transportation using different types of vehicles. For example, in Kenya the *Lipa-na-Mpesa* system of electronic are payment using mobile phones through *Mpesa* service is used in mini-bus taxis known as *Matatu.* Internet based computer platforms and mobile applications can be used for both the bus and taxi systems. Some devices, however, depending on the size can be used for both or either one of the modes. Table 4 shows ICTs and vehicles they would best suit, most of the technologies can be used for both buses and taxis.

The extent to which envisaged integration is to be implemented dictates the adaptation process. Basic technologies the stakeholders consider important have the possibility of being adopted faster depending on their benefits or expected incentives. Taxi owners, are more likely to install security devices than those entertaining passengers or

Table 4. Proposed ICT solutions and type of vehicle they would best fit

ICT solutions	Bus	Taxi
CCTV	x	x
Cell phones		
GPS		
Games and music screens		x
Seat detectors		x
Wi-Fi	x	x
Map and route display screens	x	x
SMS notifications	x	x
Information display screens	x	x
Online ticketing platforms	x	x
Countdown timers	x	x
Vehicle speed monitors	x	x
Tracking devices	x	x
Buzzers	x	x
Bluetooth kits for drivers	x	x

for fare payment if they are going to spend and not benefit from it. As the government has limited control over them, there are also limitations to what they can be directed to do or install in their vehicles. Bus service can accommodate a much more diverse ICT integrated public transportation system including vast technologies and applications. As long as it depends on government subsidy contracts, government policies can be aligned to include ICT in the services in order to be eligible for the contracts. Failure of the TRP to fully address and implement the ICT integration in the taxi industry proves that the uncontrollable nature of taxis may result in slow adoption and adaptation of ICT even with government policies regulating the process.

6 Summary

The differences in buses and taxis as modes of public transportation and their possible uptake of ICT solutions is determined by finance, management, operating environment, political involvement and stakeholder acceptance. Public transportation is South Africa was said to be "fragmented and not coordinated" (Oxford 2013), especially the taxi industry which has operations controlled by the taxi owners and drivers. These operations include payment methods, structural organisation, facilitating and adhering to policies; informal or unlawful taxi operations are also common. South Africa has illegal minibus/combi taxi operating cartels (Cervero and Golub 2011) and they account for large percentages of public transportation supply. van Ryneveld (2008), stated that mini bus taxis had to be formalised and technical and financial assistance would be offered to improve their economic viability. Bus services on the other hand as they are financially supported by the government, are relatively well organised, have clear structures and have to adhere to rules set forth by the government. Bus service operations are more corporately run compared to taxis, therefore executive decisions encouraging the use of ICT or putting ICTs in place are likely to lead to acceptance and successful integration.

The government's involvement in the transportation system, especially the bus service could be used to promote use, awareness for ICT and the adaptation process. The government using it position to update policies and ensure that they are implemented, or create new laws and legislature influencing and promoting quicker and uneventful integration. Ensuring that all government-subsidised vehicles have all relevant or suitable technologies, standardisation of such policies. Even though (Idongesit and Knud 2014) stress the point that for successful deployment of ICT in transportation there shouldn't be any cultural or political interference on ICT development plans, but currently in the Free State setting, government oversees the system.

In the case of South Africa, acceptance is also influenced by political agendas: The TRP's EMS factor would also have been successful had there been acceptance by stakeholders, in this case Taxi Associations and taxi owners. The policy provided for ICTs to be included in the vehicles, had this been adopted, most of the new Quantam taxis would have EMS components installed. The lack of acceptance was also as a result of the taxi industry's assumptions of the government taking over control of the service. Lack of consultation before decision making by the government about the Taxi Recapitalisation Programme prompted the rejection (Magubane and Manicom 2003).

7 Conclusions

ICTs are introduced to the system under assumptions that they will improve the system and the possible uptake by the stakeholders is overlooked. Prior to deployment, basic requirements such as user adoption and preparation of ICT in transportation under new circumstances before implementation should be addressed (Idongesit and Knud 2014).

The bus services in the current state of public transportation system are more favoured by the possibility of ICT integration, considering the financial backing of the government, the much more coordinated environment they operate in. Bus services that are operated by one service provider are likely to adapt smoothly and faster as compared to the taxi industry which is run by different taxi owners and Taxi associations.

Because of the government subsidy, buses would afford running at all times to meet customer demand without losing any profits. On the other hand, if taxis where to adopt demand responsive technologies, they would be affected financially as they would lose a lot of revenue. Buses are more coordinated as compared to taxis, which means any ICT integration which takes place would be a smoother transition for buses as compared to taxis.

Even though mini-bus taxis are the dominating method of public transportation (Walters 2013), they might be difficult to coordinate and integrate based on their traditionally unsystematic environment. Mini bus services use 16 seater mini-buses that operate short and long distance trips within the province and even operates in the remotest areas of the province, making taxis much more accessible as compared to buses. Traditional mini-bus taxis operate from different taxi ranks while buses operate from one bus rank. To coordinate and integrate taxis, an orderly solution has to be introduced, in the form of laws or incentives for cooperation; dispatching taxis would then be well coordinated, operating in real time. If the current bus system were to be integrated, it would be easier to manage and control the vehicles systematically.

The results indicated that the people are willing to continue using public transportation after integration; this is a sign that people are willing to accept ICT in their daily public transportation trips. Humans change their behaviour when systems around them change, therefore people are likely to adapt to what the technologies require of them. Planning of public transportation is based on the general needs of all travellers (Waara et al. 2015), this means that the technologies deployed into the public transportation system should satisfy the needs of everyone, also to be taken into consideration is that sophisticated technologies alone will not determine successful deployment of ICT.

In conclusion, slower adaptation of ICT into the public transportation system is evident for the taxi industry compared to the bus industry. The bus system may be faster to adopt and adapt to an ICT integrated public transportation system. The government should be willing to assist in accelerating ICT in the system, through either its policies or financial backing. Establishing policies and not seeing them through implementation hinders uptake by the taxi industry, had the Taxi Recapitalisation Programme been implemented to completion, all vehicles would have up-to-date technologies. Operational issues such as fare collections may be reduced through a centralised electronic fare payment system which would involve mini-bus taxis and buses using the same system (co-modality). This however, would work well with both these operations publicly subsidised and with no informal taxi operations. This will also reduce focus on one mode by the government; this would then allow for all modes of public transportation to be equally considered for funding and subsidiary opportunities.

References

Cervero, R., Golub, A.: Informal public transport: a global perspective. In: Urban Transport in the Developing World: A Handbook of Policy and Practice, pp. 488–518 (2011)
City of Cape Town: Budget 2014–2015. Annexure 19- Fares Policy for Contracted, Road-based Public Transport. Cape Town: City of Cape Town (2015)
Davis, F., Bagozzi, R., Warshaw, P.: User acceptance of computer technology: a comparison of two theoretical models. Manag. Sci. **35**(8), 982–1003 (1989)
Department of Transport. National Land Transport Act (2000)
Geenhuizen, M., Thissen, W.: Uncertainty and intelligent transport systems: implications for policy. IJTPM **2**(1), 5 (2002)
Idongesit, W., Knud, E.: The African mobile story, p. 218. River publishers, Aalborg (2014)
Magubane, N., Manicom, D.: Taxi owners' perceptions of the government's recapitalisation scheme: a case study of taxi owners in Pietermaritzburg. In: 22nd Southern African Transportation Conference, 14–16 July 2003, Pretoria, South Africa (2003)
Ministry of Transport, Public Works and Water Management: Public Transport in the Netherlands. Ministry of Transport, Public Works and Water Management, Den Haag (2010)
Mulupi, D.: Moving towards a cashless public transport system in Kenya. How We Made It In Africa – Insight into business in Africa (28 January 2015). http://www.howwemadeitinafrica.com/movingtowardsacashlesspublictransportsysteminkenya/46451/. Accessed 18 May 2015
NTT DATA: Public Transportation Shifts Gears with Smart Card Technology (2015). www.nttdata.com/americas. Accessed 01 June 2015
Regan, M., Horberry, T., Stevens, A.: Driver acceptance of new technology (2014)
Statistics South Africa: Census 2011 Statistical release. Statistics South Africa (2012)
Statistics South Africa: Statistical release P0320: National Household Travel Survey. Statistics South Africa, Pretoria (2014)
Transport for London: An accessible bus journey. [Video] (2012). https://www.youtube.com/watch?t=26&v=fj07TEsRbNk. Accessed 29 May 2015
Toba, L., Campbell, M., Schoeman, D., Lesia, P.: A critical examination of public transport: a case study of mangaung metropolitan municipality, South Africa. In: 48th ISOCARP Congress 2012. MLM Public Transport (2012)
Van Exel, N., Rietveld, P.: Perceptions of public transport travel time and their effect on choice-sets among car drivers. J. Transp. Land Use **2**(3), 75–86 (2010)
van Ryneveld, P.: 15 Year Review of Public Transport in South Africa with emphasis onmetropolitan areas (2008). http://www.infrastructuredialogue.co.za/wp-content/uploads/2013/05/MPT_Public_Transport_15yr.pdf

Waara, N., Brundell-Freij, K., Risser, R., Ståhl, A.: Feasible provision of targeted traveler information in public transportation: segmentation based on functional limitations. Transp. Res. Part A: Policy Pract. **74**, 164–173 (2015)

Walters, J.: Overview of public transport policy developments in South Africa. Res. Transp. Econ. **39**(1), 34–45 (2013)

Oxford, T.: The state of SA's public transport. Mail & Guardian. http://mg.co.za/article/2013-10-04-00-the-state-of-sas-public-transport. Accessed 17 Aug 2016

Social Sustainability Assessment of Groundwater Resources in Hanoi, Vietnam by a Simple AHP Approach

Nuong Thi Bui[1(✉)], Akira Kawamura[1], Hideo Amaguchi[1], Duong Du Bui[2], and Ngoc Tu Truong[3]

[1] Department of Civil and Environmental Engineering, Tokyo Metropolitan University, Tokyo, Japan
buithinuong@gmail.com

[2] Water Resources Monitoring Department, National Center for Water Resources Planning and Investigation, Hanoi, Vietnam

[3] New World Institute of Biotechnology, East China University of Science and Technology, Shanghai, China

Abstract. In Hanoi, Vietnam, the people heavily depend on groundwater resources for both domestic and industrial uses due to the seriously polluted problems occurred in most of the river streams system. Considering the social dimension in the context of sustainable development is a key issue in groundwater resources management. The main object of this study is to assess the social sustainability of groundwater resources in Hanoi by applying a simple Analytical Hierarchy Process (AHP) approach. By applying the simple AHP, the most important task is to comprehensively study the current social situation and actual problems to define the appropriate foremost components contributing to the sustainability goal. In this study, we successfully not only select the appropriate list of three main social sustainability aspects including the quantity, quality and management aspects and the twelve core social sustainability indicators but also appropriately assessing the sustainability of the groundwater resources from the social point of view. By gathering the data from trusted sources, the groundwater quantity, quality, and management aspects are socially assessed at a good level of 0.65, a good level of 0.61 and a slightly acceptable level of 0.46, respectively. The composite social sustainability index, *S*, is assessed at a strongly acceptable level of 0.57, indicating that Hanoi's groundwater resource is socially acceptable. The results of SSA are not only sensitive to data availability but also the definition of the indicators. The linear relationships do not reflect the actual situation well and the non-linear ones could be used to replace these sustainability scales of the actual values. The social sustainability indices are evaluated at the moderately high values, improbably reflecting the current problems in the target area. The study then points out the main research gaps and suggests the ways to improving social sustainability assessment for the better groundwater resource development.

J. Calautit et al. (eds.), *Towards Sustainable Cities in Asia and the Middle East*, Sustainable Civil Infrastructures, DOI 10.1007/978-3-319-61645-2_7

1 Introduction

1.1 Groundwater Sustainability Issues in Hanoi, Vietnam

Sustainable development has been defined as a process that "meets the needs of the present without compromising the ability of future generations to meet their own needs" (Brundtland 1987). This concept has nowadays become one of the critical global issues for humankind and all application fields. In terms of water resources management, ASCE (1998) defined that "*sustainable water resource systems are those designed and managed fully contribute to the objectives of society, now and in the future, maintaining their ecological, environmental, and hydrological integrity*". The proper management of water resources is very important to ensure a sustainable socioeconomic development of every country all over the world (Hutton and Bartram 2008; UNESCO 2009). Specifically, in terms of groundwater resources management, 'groundwater *sustainability*' may refer to the way of development and use of groundwater resource, in which the resource can be preserved for an indefinite time without causing any adverse eco-environmental and social consequences (Alley *et al.* 1999). Put simply, sufficient quantity and quality groundwater at acceptable prices should be available to meet social demands for domestic, industrial, agricultural, environmental purposes of the region now and in the future without causing the environment degradations such as land subsidence, saltwater intrusion, and so on (Plate 1993). Since groundwater resources play a key role in public water supplies around the world and the amount of groundwater abstraction has been rapidly and continuously increasing, achieving sustainable management of groundwater resources is one of the essential objectives for the future of countries (Mende *et al.* 2007).

In Hanoi, Vietnam, the river-streams system is pretty dense, but most of the main rivers and lakes are seriously polluted (Tong 2008) due to the discharge of industrial, agricultural, aqua-cultural and domestic waste to the water bodies without treatment. As described by United Nations University (2015), the water quality in the rivers inner city Hanoi is organic pollution, eutrophication, microbial contamination at a high level and signs of heavy metal pollution including Fe and Cr^{6+}. Specifically, regarding the surface water quality in the rivers, dissolved oxygen concentration (DO) is low, below the level of development of some species (> 4); BOD5 and COD contents beyond the allowable limit of National Technical Regulation (NTR) from 1.1 to 20 and from 1.1–10 times, respectively, indicating that the organic pollution level is high; and about the heavy metal pollution, there are a considerable percentage of the samplings including Fe and Cr^{6+} concentrations exceeding NTR from 1.1 to 20 and from 1.3 to 4 times, respectively. That is why groundwater resources are the main water supply sources for the local residents. Recently, up to 93% of the public water supply (not private) is provided by 13 main treatment water plants which are extracting groundwater as their main sources and the Song Da water plant, which obtains its water from the Da River, contributes about 7%, equivalent to 43,000 m^3/day only (Hanoi Water Company 2013). The resource is the target area addressing the sustainability issues because the use and development of the main water supply systems obviously have a big effect on economic development, environmental protection, and social needs. In terms of quantity, there have been a number of our previous Hanoi-targeted studies comprehensively investigated groundwater potential resources (Bui *et al.* 2012a) and level trends in Hanoi (Bui *et al.* 2012b); presented the current situation of groundwater abstraction from sustainability point of view (Bui *et al.* 2016a). The rapid exploitation of the groundwater without an appropriate management system has caused a series of adverse impacts such as drying up of shallow wells, decline of groundwater level and land subsidence (Tong 2008; Bui *et al.* 2012a). In terms of quality, furthermore, we have studied about hydrogeochemical assessment of groundwater quality during dry and rainy seasons for the two main aquifers (Nguyen *et al.* 2015a); clustered hydrogeochemical groundwater data comprising major ions to investigate the seasonal and spatial hydrogeochemical characteristics of groundwater in the Pleistocene confined aquifer of the Red River Delta where Hanoi is located (Nguyen *et al.* 2015b). However, there have been very few such studies dealing with the integrated sustainability assessment of the groundwater resources, in which we (Bui *et al.* 2016b) mainly focused on sustainability assessment from the environmental point of view as one of the very few examples newly investigated in this area. To date, the integrated social sustainability assessment has not been carried out yet even though the human wellbeing and the public supports are essential for successful implementations of any water-related projects and policies.

1.2 Social Sustainability Assessment of Groundwater Resources

In order to meet their ever increasing needs, humankind continuously and progressively extracts and exploits natural resources. In terms of groundwater resources, people withdraw the natural resources for their daily life activities, develop the advanced techniques to more efficiently abstracting the resources and discharge the wastewater sources after use to make the groundwater quality anthropogenically degradable beside the natural causes. One day, the resources are over-exploited because of the continuously increasing social needs; drying up of shallow wells, a decline of groundwater level and even land subsidence have probably occurred; the resource becomes polluted and adversely affects to the human well beings. To adapt these situations, protecting the natural groundwater resources and using the various treatment methods to make it cleaner appropriately are used. While sustainable development is a concept composing of the environmental, economic and social criteria; it is acknowledge that social dimension has received less consideration in comparison to the other criteria (Vallance *et al.* 2011). The other important thing is that public supports are essential for successful implementations of any water-related projects and policies; and there has been a need to better understand public attitudes toward water resource management (Randolph and Troy 2008; Dolnicar and Hurlimann *et al.* 2011). Therefore, considering the social criteria in the context of sustainable development is a key issue in the groundwater resources development.

Social sustainability is defined as "*ensuring the sustenance of the diverse social relations that exist in healthy communities, creating the physical, cultural and social places that support wellbeing and a sense of community involves a process of engagement with the people who inhabit those places.*"(Palich, and Edmonds 2013). So that which methods are used to measure the social sustainability appropriately. To this end, Multi-Criteria Decision Making (MCDM) methods have been considered proper for sustainability assessment (Boggia and Cortina 2010). In such MCDM applications, principles are general conditions for achieving sustainability, which may be seen as the ultimate goal. Hence, sustainability should be formulated as a general objective to be achieved. The goal may be reviewed as the three fundamental pillars of sustainability including environmental, social, and economic criteria while sustainability indicators provide measures of change in the criteria over time. The purpose of sustainability indicators for an industry is to provide information on how it contributes to sustainable development (Azapagic 2004). The indicators should be easy to measure, cost effective, accommodate changing conditions, scientifically sound, and based on functional ecological relationships (Worrall *et al.* 2009). In this way, indicators can provide information for policy makers and aid in decision making (Niemeijer and de Groot 2008). Therefore, finding out the important social sustainability indicators is one of the main tasks for social sustainability assessment. What are the main components presenting for social sustainability achievements since the social sustainability indicators are context dependent and need to reflect the nature and requirements of the local community (McKenzie 2004)? Chan *et al.* (2008) found out the six critical factors for improving the social sustainability of the urban renewal project by collecting the results from a questionnaire given to planners, property development managers, and local citizens in Hong Kong. The six factors are listed up as "*satisfaction of welfare requirements*", "*conservation of resources and the surroundings*", "*creation of harmonious living environment*", "*provisions facilitating daily life operations*", "*form of development*" and "*availability of open spaces*". Interpretations of those factors for groundwater resources development, the social sustainability indicators should reflect the facilities of the water supply system for the local residents, the social satisfactions of groundwater quantity and quality as well, and the effects of the quantity and quality on human safety and health. More importantly, in terms of water supply system management, the local government have a vital role in driving the society toward sustainable development. They should manage the appropriate national/provincial/local budget to maintain the system, raise public awareness of water resources conservation and security and also understand the needs of local residents to make sure their policies and strategies effective.

Similarly in Hanoi, while there have been a series of our previous Hanoi-targeted studies and others focusing on the groundwater quantity (Bui *et al.* 2012a; Bui *et al.* 2016a, b), decline of groundwater level (Bui *et al.* 2012b), groundwater quality such as groundwater arsenic, coliform, and nitrogen contaminated situations (Berg *et al.* 2001; 2007; Bui *et al.* 2007; Bui *et al.* 2010; Nguyen *et al.* 2012; Nguyen *et al.* 2015a, b), etc., there have been a few studies regarding how these changes adversely affecting the human safety and health, and almost no studies mentioned the relative measurement of the social sustainability. For instance, Berg *et al.* (2001; 2007) presented a threat of arsenic contamination of the Red River alluvial tract in the city of Hanoi. The research indicated that several million people consuming untreated groundwater sources might be at a considerable risk of chronic arsenic poisoning. We (Bui *et al.* (2007; 2010)) presented the prevalence of arsenic contamination in both two main aquifers in Hatay province (now is combined into Hanoi) and its health effects on the community with the supports of GIS and Mathematical model. Recently, Agusa *et al.* (2014) shows the another evidence of human exposure to arsenic from drinking water in three rural districts of Hanoi via investigating and analyzing the human hair and urine samples as the bio-indicators for arsenic exposure. However, these limited number of papers mainly focus on the warnings of health and safety risks caused by the arsenic contamination while the effects of the water resources on society broadly consist of not only that but also other contamination agents such as coliform and nitrogen; the social satisfactions of groundwater quantity, quality, facilities, the effects on human wellbeing and management in terms of sustainability.

1.3 Analytical Hierarchy Process (AHP) Approach

The MCDM methods have been considered as a proper approach for sustainability assessment (Boggia and Cortina 2010). AHP is one of the most popular and powerful MCDM methods (Saaty 2000) because it can help decision makers to cope with multifaceted and unstructured problems such as environment, economic and social. The AHP has been used as a widespread decision-making analysis tool for modeling unstructured problems in areas such as political, economic, social, and management sciences (Yu 2002). AHP has been utilized in a variety of sustainability assessment for a number of application fields including regional water resource (Sun *et al.* 2016). The main advantage of those applications is to categorize and identify the foremost components (aspects and indicators) that better reflect the significant performances. The indicators have been considered as an important communication tool for policy-makers, managers and the public (Chen *et al.* 2015). However, there have been very few such studies dealing with sustainability assessment of groundwater resources, in which Chen *et al.* (2015) focusing on Hohhot Plain in China as one of the very few examples investigated in the semiarid regions where the annual precipitation is about 408 mm only. There have been no social sustainability assessment (SSA) studies carried out by this outstanding MCDM method for groundwater resources so far.

Dealing with these above-mentioned problems and in order to assess the social sustainability degree for Hanoi's groundwater resources, this paper utilizes the simple AHP to properly weighting the sustainability contribution of the components. This study develops the SSA framework in checking with the actual data's availability and assesses the sustainability from the social point of view. Based on the discussion in the social sustainability assessment (SSA), the paper provides the useful recommendations to improve SSA's performance. The results are fundamental for further integrated sustainability assessment for groundwater resources in Hanoi.

2 Study Area

Figure 1 shows the geographical location and the main rivers and lakes of Hanoi. Hanoi is located in the north-eastern part of Vietnam with the area of 3324.5 km^2. The population of more than 7.2 million (as of 2015) accounts for almost 10% of Vietnam in total with the highest density of more than 2 thousands people/km^2 (General Statistic Office of Vietnam 2015). Hanoi belongs to the tropical monsoonal area with two distinctive seasons in the year, the rainy season from May to October and the dry season from November to April of the following year. The annual average rainfall is about 1,600 mm, the average humidity is about 80%, and the average temperature is around 24.3°C. Evaporation is quite high with an annual average of 933 mm. Hanoi also has a dense river network (density of 0.7 km/km^2) and is mainly supported by Red River, one of two biggest river systems, with the basin areas of approximately 155,000 km^2. However, the fast economic development and socialization and urbanization have put pretty much pressure on the river basin environment. This surface water system is recently seriously polluted due to a large amount discharge of untreated sewage, industrial waste, and garbage, in which, the lakes are significantly polluted (Tong 2008). The water quality in the rivers is organic pollution, eutrophication, microbial contamination at a high level and signs of heavy metal pollution including Fe and Cr6+ (United Nations University 2015). That is the main reason why the groundwater resources have become the most important water supply for the local inhabitants.

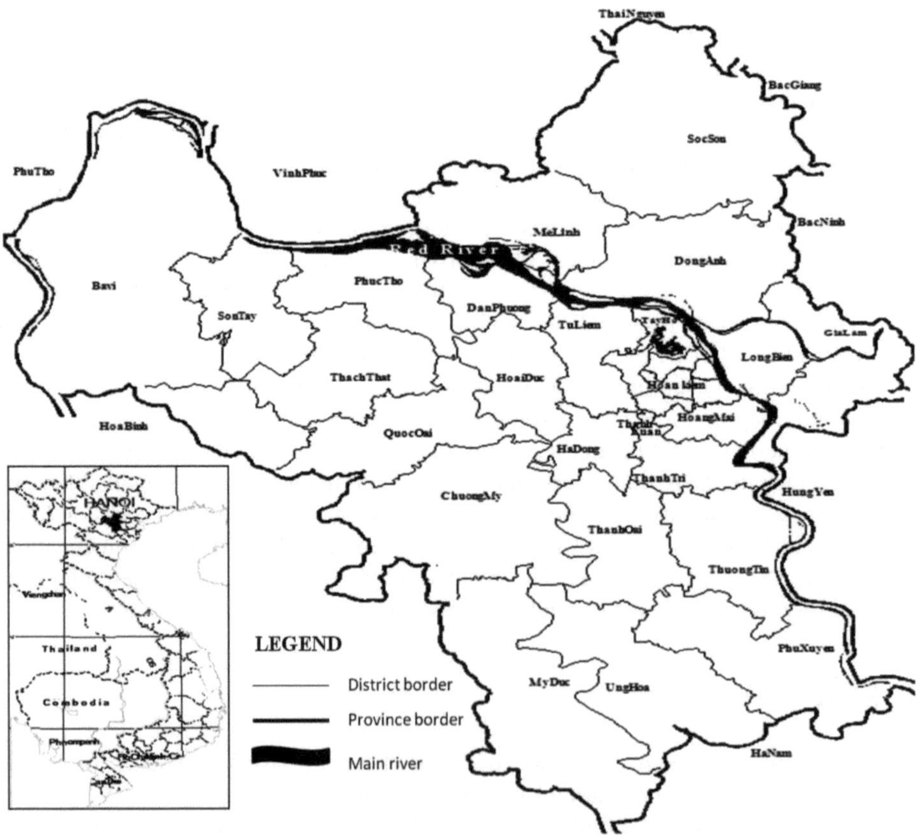

Fig. 1. Study area and main rivers

3 Methodology

Established in the 1970s by Saaty, AHP is one of the most powerful and popular multi-criteria decision-making methods dealing with multifaceted and unstructured problems such as political, economic, social, and management sciences (Yu 2002). To apply AHP, at first step, decision makers need to study the current situation of the complex multiple criteria decision problems (for example, social sustainability) to create the appropriate hierarchy by breaking down it into its aspects and the corresponding indicators in each aspect. Secondly, the relative contribution of the indicator and aspect to the final goals are defined by series of the consistent judgments from the experts. Based on the results of AHP, decision makers could see which aspects and indicators should/shouldn't be improved to enhance the sustainability performance.

3.1 SSA for Groundwater Resources by the Simple AHP

Figure 2 shows the basic evaluation steps in SSA evaluation based on the simple AHP proposed by Bui *et al.* (2016b).

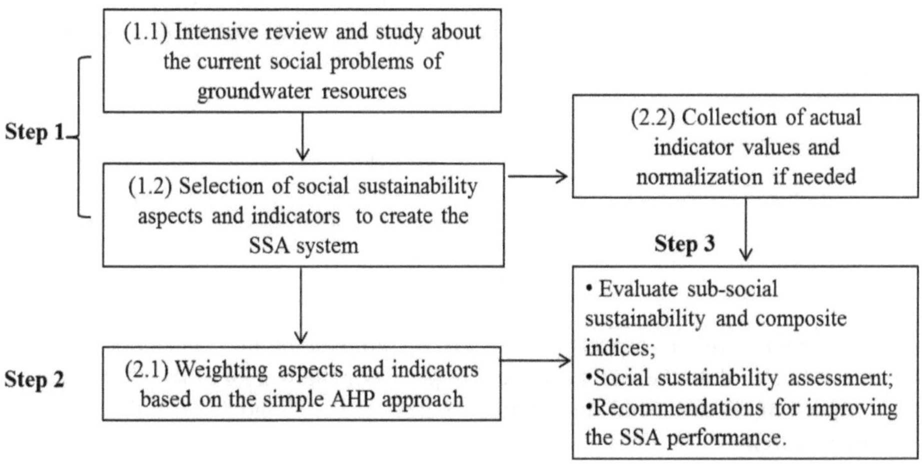

Fig. 2. Social sustainability assessment (SSA) based on the simple AHP approach.

Step1: *Establishing the multiple-level hierarchy*

Decision makers need to study the current situation of the complex multiple criteria decision problems (social sustainability) to create the appropriate hierarchy by breaking down it into its sustainability aspects and the corresponding sustainability indicators in each sustainability aspect.

Step 2:

- The relative contribution of each aspect and indicator to the corresponding goals are defined by the expert's comparison judgments in the standard AHP. In the standard

AHP, the performance ratings and the weights of the attributes result from series of pairwise comparison judgments between two attributes at the same level of the hierarchy, which are given in crisp numbers from 1 to 9. Practically, it is difficult to extract precise data pertaining to measurement factors because human preferences normally include a degree of uncertainty and it is unrealistic to expect that decision-makers have either complete information or a full understanding of all aspects of the problem (Boender *et al.* 1989). Therefore, these judgments are the 'unconfident pair-wise comparison judgment. The second step is considered as the most practically time-consuming and complicated due to several reasons such as: finding the appropriate experts; waiting for their big efforts to make the large series of the unconfident pair-wise comparison judgments; making these judgments again and again until they become acceptably consistent (Saaty 2000). In Hanoi, however, there are no such complicated surveys carried out to consult the expert's opinions regarding groundwater sustainability assessment. In order to cope with above-mentioned problems, we aimed to modify the standard AHP to make it simple in the way of properly weighting the contribution of each sustainability component to the final goal.

- A collection of actual data values for all the indicators.

Step 3: The social sustainability index (SA_i) of the ith aspect is evaluated by the Eq. (1) and the composite index for social sustainability assessment, S, is assessed by the Eq. (2).

$$SA_i = \sum_{j=1}^{N_i} W_{ij} * SI_{ij} \tag{1}$$

$$S = \sum_{i=1}^{N} WA_i * SA_i \tag{2}$$

where N: number of the aspects; $i = 1\ldots N$; N_i: number of the indicators in the ith aspect; $j = 1\ldots N_i$; WA_i: the weight of the ith aspect; W_{ij}: the weight of the jth indicator in the ith aspect; SI_{ij} is the social sustainability index for the jth indicator in the ith aspect, with the constrains:

$$0 \leq WA_i, W_{ij} \leq 1; \tag{3}$$

$$\sum_{i=1}^{N} WA_i = 1; \sum_{j=1}^{N_i} W_{ij} = 1; \tag{4}$$

So that naturally, those indices are in a range of (0–1).

- In this research, the relative contribution of each aspect and indicator to the corresponding goals are defined by the simple AHP. The simple AHP is an approach in which the weighting process by the function of a number of the sustainability aspects and the indicators is used to replace the ones by the expert's comparison judgments in the standard AHP. In the simple AHP, particularly, once these foremost components are decided, their weights can be made automatically by the following Eqs. (5) and (6):

$$W_{ij} = \frac{1}{N_i} \tag{5}$$

$$WA_i = \frac{1}{N} \tag{6}$$

Sustainability Scale: the groundwater sustainability can be classified into five classes on a scale of 0–1:

Very poor level: from 0 to 0.2; poor level: from 0.2 to 0.4; slightly acceptable level: from 0.4 to 0.45; acceptable level: from 0.45 to 0.55; strongly acceptable level: from 0.55 to 0.6; good level: from 0.6 to 0.8 and excellent level: from 0.8 to 1.0. The highest scores for sustainability is 1, and the lowest one is 0.

4 SSA Application to Hanoi Groundwater Resources

UNESCO/IAEA/IAH Working Group is the group first trying to define the sustainability of groundwater resources which follows the DPSIR (Driving forces, Pressures, State, Impacts and societal Response) framework (Vrba and Lipponen 2007). Those indicators are related to generally groundwater situations and can be used as the broad guideline to establish the indicators. However, those indicators are the independent contributions to the sustainability from different points of view; in other words, these indicators assess the sustainability not in a common system. Therefore, from a specific point of view, it is obviously difficult for specific applications of those indicators into other areas. Since AHP has been considered as one of the powerful indicator-based approaches in the literature review, this paper is the first effort in building up the list of sustainability aspects and the corresponding indicators in each aspect for groundwater resources in a common system based on the AHP concept. By using the AHP approach, the components (aspects and indicators) for groundwater sustainability should be created with the foundation knowledge of the current situations, actual problems occurred and the expected goal (Chen *et al.* 2015), the aspects need to covered all the dimensions of the final goal concept and the corresponding indicators are the smallest units in the system and physically measurable. The more complex indicators system can be developed if the more actual data are available.

Table 1 shows the main sustainability aspects and indicators built up for the first time for the valuable groundwater resources from the social point of view and Hanoi situation is a case study. Three proposed aspects are quantity, quality, and management with the following basic reasons in this study area. Even the excessive groundwater abstraction has caused serious groundwater-level declines in the central and south parts of Hanoi, there is still some information of insufficient water use reported in Vietnam's newspapers. In 2016, approximately two days per month the urban districts having no water supplied from the public water supply companies (Hanoi Water Limited

Company 2016). This insufficient water use apparently adversely affects to the daily life routines of the local residents, especially in the summer season when the temperature even reaches 45°C in some central areas. About the quality, the untreated groundwater resources are reported as arsenic, nitrogen, and coliform contaminated by both natural and anthropogenic causes in the literature review. More dangerously, the contamination is still existed in the bottled water and treated water supply (Craig *et al.* 2013). It is not surprising that these problems adversely impact on local community in both short and long-term exposures. To face to these problems, how the local government manages for driving Hanoi towards sustainable development from the social point of view. Therefore, in this study, from the social sustainability point of view, the social considerations of the groundwater quantity, quality and management are considered as three main social sustainability aspects.

Table 1. SSA framework for groundwater resources in Hanoi, Vietnam

Aspect	Indicator	Definitions
Quantity (SA_1)	SI_{11}	Ratio of the number of residents who can access water for living to the total population in the study area
	SI_{12}	Ratio of the number of days per month, local residents having sufficient water use in the urban area
	SI_{13}	Ratio of the number of hours per day, local residents having sufficient water use in the urban area
Quality (SA_2)	SI_{21}	Ratio of number of residents who use the groundwater with no arsenic contamination to the total population
	SI_{22}	Ratio of number of residents who use the groundwater resources with no nitrogen contamination to the total population
	SI_{23}	Ratio of number of residents who use the groundwater resources with no coliform contamination to the total population
	SI_{24}	Ratio of number of residents who have no water related diseases to the total population
Management (SA_3)	SI_{31}	Ratio of the number of people who can access to the public water supply system to the total population
	SI_{32}	Ratio of the government budget allocated in integrated water resources management (IWRM) to the budget need for maintaining the water supply system
	SI_{33}	Ratio of the number of good responds from local residents to the water supply management of the local government
	SI_{34}	Ratio of number of respondents who are willing to participate (WTPa) in any water conservation and protection activities to the total population
	SI_{35}	Ratio of number of respondents who are willing to pay (WTP) for improve the water supply system for wellbeing to the total population

Regarding the quantity aspect, we focus on how much social satisfaction of water usage in terms of the quantity since groundwater is the main water supply. The terms of "satisfaction" and/or "sufficient water use" are difficult to define. Depending on social needs and situation, the amount considered as "enough" is totally different. As one of the developing countries, we define that "sufficient water use" means people can access and have water for the basic daily activities. As guided by the UNESCO/IAEA/IAH Working Group, the indicators are defined as the ratios between numbers of residents who have insufficient water use to the population in a quantitative aspect. However, in this study, the sustainability indicators indicate that the bigger values of the indicators are, the better contribution can be made to the final social sustainability goal. That is the reason why we define as ratios of the number of residents having sufficient water use to the total population. More specifically, our indicators can relatively measure how many days per month and how much time in 24 h of the no-water-provided day, the residents can have the access water use from the public water supply system. By these definitions, the socially sustainability contributions of the indicators are maximized at one if anyone has sufficient water use.

Regarding the quality aspect, we focus on the risk of water consumption for the residents as the answer to the "how many people who are using the contaminated groundwater resources for living?" question and the water-related-diseases situation in Hanoi. In Vietnam, up to 80% of diseases in Vietnam are caused by polluted water resources, said the Ministry of Natural Resources and Environment, and about six million Vietnamese people have contracted one of six water-related diseases (such as bacterial diarrhea, hepatitis A, typhoid fever, dengue fever, malaria, and Japanese encephalitis) over the four-year period. In this study area, there are three main pollution concerns such as the arsenic, nitrogen and coliform contaminated groundwater, thus the indicators are defined as the ratios between the numbers of residents who are probably not affected due to living in the no-contaminated areas to the total population. The indicators regarding arsenic risk, SI_{21}, by the contaminated groundwater resources. Similarly, the SI_{22} and SI_{23} are defined as the ratios between the numbers of residents who are probably not affected due to living in the no-nitrogen/coliform contaminated areas to the total population. The indicator SI_{24}, furthermore, considers to the water-related diseases of the residents due to the contaminated groundwater consumption. By these definitions, the social sustainability contributions of the indicators are maximized at one if there is no one using the polluted water resources and minimized at zero if all the water supply sources are polluted.

Regarding the management aspect, this paper considers how the local government manages and improves the water supply system for better use and how the residents respond to the management by their willingness to pay (WTP) for improving the public system. The indicator SI_{31} mainly considers the sufficient water supply facilities and how much percentage of the residents who can access the water supply system piles. The indicator SI_{32} presents the sufficient budget allocation in integrated water resources management (IWRM), compared to the expected budget needed for maintaining the system. These two important indicators show how much the government cares about the water supply system in their development strategy. However, it is obviously missing if we do not care about what and how the local residents say about the management. In fact, as resulted in our previous pilot study in Hanoi City about public awareness, attitudes and behaviour towards water management issues, there are up to 85% of the respondents are not actively participated in any water conservation and protection groups (Bui *et al.* 2014) even though there are about 56% of local residents who are willing to contribute financial supports to improve water quality in general. So that the big question for the government is how to raise the very poor understanding of water use and water resources for the local residents. This thing could help the decision makers evaluate their performance and improve it to make it much more closed to the actual social needs. The indicator SI_{33}, SI_{34} and SI_{35} are mainly about the response from local residents, how much interest on the water related programs and how much their willingness to pay for improving the water supply system.

In this study, the three social sustainability aspects and twelve indicators in Table 1, are considered as the core components presenting the actual social scenarios of groundwater resources in Hanoi. The more complex system can be developed if the more actual data are available and the specifically different point of view.

5 Results

This study calculates the actual values of the aforementioned social sustainability indicators (SI_{ij}) by gathering the necessary data from the government database, Ministry of Health Portal and Ministry of Natural Resources and Environment, Vietnam.

Table 2 shows the twelve indicators in three sustainability aspects with their formula, variables used and the explanations. Table 3 shows the results with weights defined by the simple AHP approach. The groundwater quantity index (SA_1) is assessed at a socially good level of 0.65; the quality index (SA_2) is relatively assessed at a

Table 2. Groundwater social sustainability indicators, formula, variables used and value

Indicator	Formula	Variables used/Explanation	Value
SI_{11}	(the number of residents can access clean water for living)/(the total population)	About 70% local residents can access clean water for living in the special-level city (Ministry of Health Portal 2009). About 36.68% local residents in the rural areas of Hanoi can access clean water for living in 2014, said Tran Xuan Viet, the Vice Chairman of Hanoi City Committee. Hanoi is currently extended and includes more than 3.2 million and 3.9 million residents living in the urban districts and sub-urban districts, respectively. We take the calculation for SI_{11}: (0.7*3.2 + 0.37*3.9) million/7.1 million = 0.52	0.52
SI_{12}	(the number of days, local residents having sufficient water use)/(30 days)	Approximately two days per month in 2016, the urban districts having no water supplied from the public water supply companies (Hanoi Water Limited Company 2016)	0.93
SI_{13}	(the number of hours in the no-water-supplied day, local residents having sufficient water use)/(24 h)	In 2016, approximately 12 daily hours per 24 h in the no-water-supplied day, the urban districts having no water supplied from the public water supply companies (Hanoi Water Limited Company 2016)	0.50

(*continued*)

Table 2. (*continued*)

Indicator	Formula	Variables used/Explanation	Value
SI_{21}	(the number of residents who use the groundwater water with no arsenic contamination)/(the total population)	Estimated that 10 millionpeople in the Red River delta are at risk of chronic arsenicpoisoning (Berg *et al.* 2007). Total population in Red River Delta is about 11 million people inhabited. We simply take this roughly estimation presenting for Hanoi.	0.09
SI_{22}	(the number of residents who use the groundwater resources with no nitrogen contamination)/(the total population)	About 43% ammonium, 15% nitrate dioxide and 12% nitrate of the water samples in Hanoi are not permissible for drinking water (Nguyen *et al.* 2012); the maximum percentage is about 43%, thus the possible largest area with no ammonium, nitrate dioxide and nitrate-contaminated groundwater could be 57%	0.57
SI_{23}	(the number of residents who use the groundwater resources with no coliform)/(the total population)	About 22% of samples in both the Hanoi aquifers have coliform values higher than the standard limit in Hanoi (Nguyen *et al.* 2012)	0.78
SI_{24}	(the number of residents who number of residents who have no water related diseases)/(the total population)	Around six million Vietnamese people have contracted one of six water-related diseases over the past four years. (http://www.ngocentre.org.vn/content/80-diseases-vietnam-caused-polluted-water-resources)	0.98
SI_{31}	(the number of people who can access to the public water supply system)/(the total population)	About 70% local residents can access clean water for living from the public water supply system in the special-level city and in the rural areas (Ministry of Health Portal 2009). About 10% local residents can access clean water from the public water supply system in the rural areas. Hanoi is currently extended and includes more than 3.2 million and 3.9 million residents living in the urban districts and sub-urban districts, respectively. We take the calculation for SI_{31}: (0.7*3.2 + 0.1*3.9) million/7.2 million = 0.36	0.37
SI_{32}	(the government budget allocated in integrated water resources management (IWRM))/(the budget need for maintaining the water supply system)	In 2006, the total budget of MARD was US$200 million of which US$1r26 million was allocated to investment for development. (Molle and Hoanh 2011)	0.63
SI_{33}	(number of good responds from local residents)/(the total population)	There are about only 6% of respondents commented that the management and propaganda methods of government departments are good, while most citizens (43%) rated as poorly managing and protecting water quality. (Bui *et al.* 2014)	0.57
SI_{34}	(number of residents who are willing to participate in any water conservation and protection activities)/(the total population)	There are about 85% of the public is not actively participated in any water conservation and protection groups. (Bui *et al.* 2014)	0.15
SI_{35}	(number of residents who are willing to pay for improve the water supply system)/(the total population)	There are about 56% of local residents who are willing to contribute financial supports to improve water quality. (Bui *et al.* 2014)	0.56

socially good level of 0.61 and the management index (SA_3) is assessed at a slightly socially acceptable level of 0.46. The composite social sustainability index, S, is about 0.57, socially assessed at strongly acceptable level.

6 Discussion

From Table 3, the groundwater quantity index (SA_1), which is assessed at a socially good level of 0.65, indicates that the community rather satisfies with the water accessibility and the amount of daily water use. The quality index (SA_2) of 0.61 indicates that the community also moderately satisfies with the quality of the groundwater resources. The management index (SA_3) of 0.46 implies that the community somewhat accepts the current policies and regulation of the government management. Consequently, the composite social sustainability index of 0.57 shows that the groundwater use and condition in Hanoi is socially strongly acceptable. Among these indices, it is effortless to see that the quality index is somehow inappropriately assessed from the sustainability point of view. The following paragraph explains the reasons why and also points out how to improve the assessment properly.

Table 3. Social sustainability assessment for Hanoi groundwater resources

Sustainability aspect	W_i	Sustainability indicator	W_{ij}	Values of SI_{ij}	SA_i	S
Quantity (SA_1)	0.333	SI_{11}	0.33	0.52	0.65	0.57
		SI_{12}	0.33	0.93		
		SI_{13}	0.33	0.50		
Quality (SA_2)	0.333	SI_{21}	0.25	0.09	0.61	
		SI_{22}	0.25	0.57		
		SI_{23}	0.25	0.78		
		SI_{24}	0.25	0.98		
Management (SA_3)	0.333	SI_{31}	0.20	0.37	0.46	
		SI_{32}	0.20	0.63		
		SI_{33}	0.20	0.57		
		SI_{34}	0.20	0.15		
		SI_{35}	0.20	0.56		

From Table 2 and Fig. 3, the quality aspect (SA_2) is assessed at the socially good level. Since there have been a series of reports regarding the serious pollution problems published, this is a not easy-to-believe assessment from the quality point of view. Because we could see that, the indicator SI_{21} in Table 2 shows that almost 90% of local residents in Red River Delta are at risk of arsenic poisoning due to the arsenic contaminated groundwater resource consumption; and the indicator SI_{24} shows that approximately 2% population have contracted one of six water-related diseases over the four-year period. Therefore, in terms of human health and safety, the sustainability scales should be appropriately changed in a more realistic way. In addition, due to the

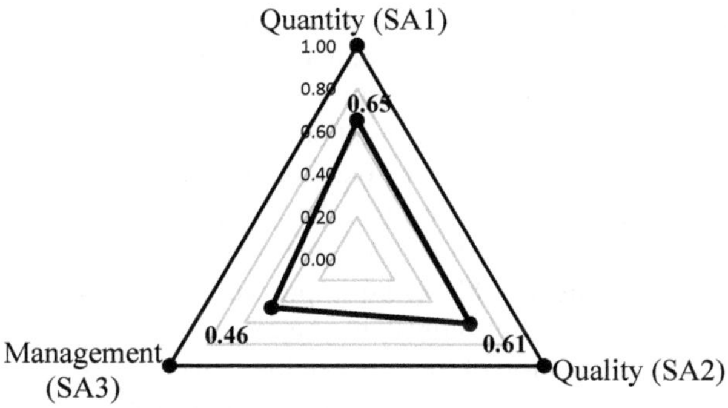

Fig. 3. Social sustainability assessment of Hanoi groundwater resources

limitations of the data availability, the indicators (SI_{22} and SI_{23}) in this study are evaluated as the same as the area ratios. These area ratio estimations do not locally reflect the actual problems appropriately due to the moderately small proportion between the contaminated areas and the total study area. Previously, Hanoi was small and the abstraction wells are mainly located in the central area which becomes the most vulnerable areas regarding groundwater over-exploitation due to the fast population growth and the rapid urbanization. Recently, since 2008, Hanoi has been largely extended; the previous area is about 15% of the current. The sustainability indicators show the general sustainability assessment in the total extended area. Therefore, the values of these indicators do not locally reflect the actual problems appropriately. Another thing, in Table 2, is that the measurement values in the description for the indicators of the second aspect were at different scattered sites in Hanoi. The way to convert the point-based measurements into the area absolutely encompasses the uncertainty and error in the calculations. However, all the groundwater samples were broadly collected over the study area. That is why to cope with the data availability; we use the assumption in which the point-based measurements are the representatives for the area. Therefore, not only the sustainability scales should be changed from social point of view but also these values of the actual data (SI_{ij}) should be validated to make the evaluation much more closely to the reality.

The results of SSA are not only sensitive with data availability but also the definition of the indicators. Regarding to quality aspect (SA_2), the indicator formulas are defined by ratios of the number of residents who use the groundwater with no contamination to the total population, inspired from the way to define the indicators of UNESCO/IAEA/IAH Working Group. For example, in the indicators SI_{22}, there are about 43% areas with nitrogen contamination of the total; the sustainability level is strongly acceptable at the value of 0.57. This sustainability scale is apparently not suitable due to the severely adverse impact of the nitrogen exposure from drinking water in general. For instance, excess the WHO standard of nitrates (50 mg L−1 (WHO 2011)) in the drinking water cause human health risks of depleting blood oxygen levels; the additional consequences are enlargement of the thyroid gland, increased incidence of cancer and birth defects, and hypertension (Forman *et al.* 1985). Such these linear relationships, therefore, do not reflect the actual situation well. In order to avoid this inappropriate scale, we could use a non-linear relationship to define the SI_{22}, and we need to make another detailed sustainability scales for this quality indicator. The sustainability could be suitably scaled in a manner, in which, (i) if the proportion of the nitrogen contaminated area to the total area is 0% and 100%, the sustainability scale converted should be 1 and 0, respectively; and (ii) if this proportion is 50%, the sustainability scale should be converted into 0.1, for instance. In general, in order to improve the sensitivity of SSA performance, the more suitable definition of the sustainability indicators is indispensable to be improved to make it closed to the actual groundwater situation and the indicators could be more helpful to the decision makers.

Regarding to the modification of the methodology, in order to cope with the limited data availability, the purpose of this study is to economically reduce the most practically time-consuming and complicated step in the standard AHP due to several reasons such as: finding the appropriate experts; waiting for their big efforts to make the large series of unconfident pair-wise comparison judgments again and again until they become acceptably consistent. To do that, we carefully build up the main sustainability aspects and indicators covering the actual situation of groundwater resources in Hanoi from the social point of view. From the simple method, once the sustainability criteria and indicators are proposed, the social sustainability of the groundwater resources can be relatively assessed when the data are available. This simple sustainability assessment thus provides a quick view of the current groundwater use status and can be applied to other areas with the similar interests. Moreover, using this simple assessment, we can relatively make the comparisons among the quantity, quality, and management aspects; as well as make the comparisons among the sustainability indicators. The purpose of these comparisons is to fairly point out the most important aspects and indicators which are needed to be highly invested in order to effectively improve social sustainability. Therefore, the simple AHP could be considered as the first test of an economically substituted approach for the standard AHP; the validation should be carried out as the future work.

7 Conclusions

The main object of this study is to assess social sustainability level for the valuable groundwater resources by applying the indicator-based approach, the simple AHP. In this study, we successfully not only create the appropriate list of three main aspects and the twelve core social sustainability indicators, appropriately presenting for groundwater situation in Hanoi, but also test the simple AHP approach in handling the limitation of data availability. The results of SSA are not only sensitive to data availability but also the definition of the indicators. The linear relationships do not reflect the actual situation well and the non-linear ones could be used to replace these sustainability scales of the actual values. The social sustainability indices are evaluated at the moderately high values, improbably reflecting the current problems in the target area. The study then points out the main research gaps and how to improve social sustainability assessment for the better groundwater resource development.

Acknowledgements. This study was carried out as a part of the research project "Study on guerrilla rainstorms, flood inundation and water pollution in metropolitan watersheds" supported by the Tokyo Metropolitan Government, Japan (represented by A. Kawamura). We would like to thank Ministry of Natural Resources and Environment (MONRE) of Vietnam for supplying the necessary field data from the earlier feasibility studies.

References

Alley, W.M., et al.: Sustainability of Ground-Water Resources U.S. Geological Survey Circular 1186, Denver, CO (1999)

American Society of Civil Engineers (ASCE): Sustainability Criteria for Water Resources Systems. Task Committee on Sustainability Criteria, Water Resources Planning and Management Division, ASCE and Working Group, UNESCO/IHP IV Project M-4.3. ASCE, Reston. Va (1998)

Asuga, T., et al.: Human exposure to arsenic from drinking water in Vietnam. Sci. Total Environ. **488–489**, 562–569 (2014)

Azapagic, A.: Developing a framework for sustainable development indicators for the mining and minerals industry. J. Cleaner Prod. **12**(6), 639–662 (2004). doi:10.1016/S0959-6526(03)00075-1

Berg, M., et al.: Arsenic contamination of groundwater and drinking water in Vietnam: a human health threat. Environ. Sci. Technol. **35**(13), 2621–2626 (2001)

Berg, M., et al.: Magnitude of arsenic pollution in the Mekong and Red river deltas — Cambodia and Vietnam. Sci. Total Environ. **372**, 413–425 (2007)

Boender, C.G.E., et al.: Multi-criteria decision analysis with fuzzy pairwise comparisons. Fuzzy Sets Syst. **29**, 133–143 (1989). doi:10.1016/0165-0114(89)90187-5

Boggia, A., Cortina, C.: Measuring sustainable development using a multi-criteria model: a case study. J. Environ. Manag. **91**, 2301–2306 (2010). doi:10.1016/j.jenvman.2010.06.009

Brundtland, G.H.: Our Common Future. Oxford University Press, Oxford (1987)

Bui, D.D., et al.: Research on the groundwater pollution and its effect on the community health in Hanoi, Vietnam with the supports of GIS and Mathematical model. In: Proceedings of the International Workshop on Bio-Medicine, Hanoi, Vietnam, 25–27 July, pp. 338–345 (2007)

Bui, D.D., et al.: Spatio-temporal analysis of recent groundwater-level trends in the Red River Delta, Vietnam. Hydrogeol. J. **20**, 1635–1650 (2012a)

Bui, D.D., et al.: Aquifer system for potential groundwater resources in Hanoi, Vietnam. Hydrol. Process. **26**, 932–946 (2012b)

Bui, D.D., et al.: Public awareness, attitudes and behaviour towards water management issues in Vietnam: a pilot study in Hanoi city. In: Proceedings of the 4th Vietnam Water Cooperation Initiative (2014)

Bui, T. N., et al.: Prevalence of Arsenic in groundwater resources in Hanoi2, Vietnam. In: SEGH 2010 International Conference and Workshops of the Society for Environmental Geochemistry and Health on Environmental Quality and Human Health, Galway, Ireland, pp. 51–52 (2010)

Bui, T.N., et al.: Sustainability assessment of groundwater abstraction in Hanoi, Vietnam. In: The 7th International Conference on Water Resources and Environment Research (ICWRER2016). Kyoto, Japan, 5th-9th June 2016 (2016a)

Bui, T.N., et al.: Environmental sustainability assessment of groundwater resources in Hanoi, Vietnam by a simple AHP approach. J. JSCE, Ser. Glob. Environ. Res. **72**(5), I137–I146 (2016b)

Chan, E., Lee, G.K.L.: Critical factors for improving social sustainability of urbanrenewal projects. Soc. Indic. Res. **85**, 243–256 (2008)
Chen, J., et al.: Improving assessment of groundwater sustainability with analytic hierarchy process and information entropy method: a case study of the Hohhot Plain. China. Environ. Earth Sci. **73**, 2353–2363 (2015)
Craig, P., et al.: Arsenic contamination in Hanoi treated water supply. Book of Abstracts of the 3rd Vietnam Water Cooperation Initiative (2013)
Dolnicar, S., et al.: What affects public acceptance of recycled and desalinated water? Water Res. **45**(2), 933–943 (2011)
Forman, D., et al.: Nitrates, nitrites and gastric cancer in Great Britain. Nature **313**(6004), 620–625 (1985). doi:10.1038/313620a0
General Statistic Office of Vietnam: Area, population and population density by province by Cities, provinces, Year and Items (2015). https://www.gso.gov.vn/default_en.aspx?tabid=774
Hanoi Water Limited Company Ltd (2016). http://www.hawacom.vn
Hutton, G., Bartram, J.: Regional and global costs of attaining the water supply and sanitation target (Target 10) of the Millennium Development Goals. World Health Organization, Geneva (2008)
McKenzie, S.: Social Sustainability. Towards Some Definitions; University of South Australia: Adelaide, Australia (2004)
Mende, A., et al.: Strategy for groundwater management in developing countries: a case study in northern Costa Rica. J. Hydrol. **334**, 109–124 (2007)
Ministry of Health Portal: National technical regulation on drinking water quality. QCVN 01:2009/BYT (2009)
Molle, F., Hoanh, C.T.: Implementing integrated river basin management in the Red River Basin, Vietnam: a solution looking for a problem? Water Policy **13**(4), 518–534 (2011). doi:10.2166/wp.2011.012
Nguyen, T.T., et al.: An overview of groundwater quality in Hanoi, Vietnam. In: Proceedings of the 39th Kanto Branch Annual Conférence of JSCE, CD-ROM, pp. II–16 (2012)
Nguyen, T.T., et al.: Hydrogeochemical assessment of groundwater quality during dry and rainy seasons for the two main aquifers in Hanoi, Vietnam. Environ. Earth Sci. **73**, 4287–4303 (2015a)
Nguyen, T.T., Kawamura, A., Tong, T.N., Nakagawa, N., Amaguchi, H., Gilbuena, R.: Clustering spatio-seasonal hydrogeochemical data using self-organizing maps for groundwater quality assessment in the Red River Delta, Vietnam. J. Hydrol. **55**, 661–673 (2015b)
Niemeijer, D., de Groot, R.: A conceptual framework for selecting environmental indicator sets. Ecol. Ind. **8**(1), 14–25 (2008). doi:10.1016/j.ecolind.2006.11.012
Palich, N., Edmonds, A.: Social sustainability: creating placesand participatory processes thatperform well for people. Environ. Des. Guide 78 NP (2013). ISSN 1442–5017
Plate, E.J.: Sustainable development of water resources: a challenge to science and engineering. Water Int. **18**(2), 84–93 (1993)
Randolph, B., Troy, P.: Attitudes to conservation and water consumption. J. Environ. Sci. Policy **11**(5), 441–455 (2008)
Saaty, T.L.: Fundamentals of Decision Making and Priority Theory. RWS Publications, Pittsburgh (2000)
Sun, S., et al.: Sustainability assessment of regional water resources under the DPSIR framework. J. Hydrol. **532**, 140–148 (2016)
Tong, T.N.: Establishing integrated water resources database for effective management in Hanoi, The final project report. Northern Division of Water Resources Planning and Investigation (In Vietnamese) (2008)

UNESCO: The United Nations World Water Development Report: Water in a Changing World. vol. 3. United Nations Education Scientific and Cultural Organization, Earthscan, London (2009)

United Nations University: The Current Status and Solutions for Urban Water Environment in Hanoi City. Water and Urban Initiative. Working Paper Series No. 05, April 2015

Yu, C.S.: AGP-AHP method for solving group decision- makingfuzzy AHP problems. Comput. Oper. Res. **29**, 1969–2001 (2002)

Vallance, S., et al.: What is social sustainability? A clarification of concepts. Geoforum **42**, 342–348 (2011)

Vrba, J., Lipponen, A.: Groundwater resources sustainability indicators. IHP-Viseries on groundwater, 14 (2007)

WHO (World Health Organization): The Fourth Edition of Guidelines for Drinking-water Quality (2011)

Worrall, R., et al.: Towards a sustainability criteria and indicators framework for legacy mine land. J. Cleaner Prod. **17**(16), 1426–1434 (2009). doi:10.1016/j.jclepro.2009.04.013

A Computational Fluid Dynamics (CFD) Study on Enhancing Green Building Performance in Dubai, UAE Using Diffuser Augmented Wind Turbines (DAWT)

Arouge Agha(✉) and Hassam N. Chaudhry

School of Energy, Geoscience, Infrastructure and Society, Heriot-Watt University, Edinburgh, UK
aa182@hw.ac.uk

Abstract. By incorporating a Diffuser around the rotor to form a Diffuser Augmented Wind Turbine (DAWT), a conventional wind turbines' performance can be significantly improved. The diffuser is able to funnel and accelerate incident air flow to the rotor thereby driving it for a higher power extraction threshold for better control and continuous periods of operation. A steady-state Computational Fluid Dynamics (CFD) study was carried out in ANSYS Fluent on a three-bladed roof-mounted DAWT on a section of the Heriot-Watt University building located in Dubai, UAE. The results displayed that the diffuser significantly improved performance on the equivalent bare wind turbine used as a baseline model. The DAWT outperformed the Horizontal Axis Wind Turbine (HAWT) with Concentrator as it induced the largest pressure drop across the rotor and the largest wind speed. The results show improved aerodynamic capabilities as wind speeds were significantly increased and better distributed across the rotor in the DAWT; it achieved a 53.8% increase in wind speed compared to the benchmark HAWT (bare wind turbine) which had 0.6 m/s at rotor. The HAWT w. Concentrator achieved a 35.1% increase on the benchmark. The maximum power augmentation achieved was 2.5 at speed of 1.3 m/s at the rotor for the DAWT. The results agree well with the work of Ohya and Karasudani (2010) where they achieved augmentation ratios between 2 and 5 and the work of Wang and Chen (2008) where an augmentation ratio of 2.2 was achieved for DAWT's. The conclusions for this study provide an example of improving building performance with renewable wind technology. In accordance with the UAE's 2030 vision to encourage "sustainability, infrastructure capacity, community planning and quality of life", the present work hopes to contribute to the mandate.

Nomenclature

a	Axial Induction Factor
A	Area
C_D	Disk Loading Coefficient
C_p	Power Coefficient
ρ	Air Density
P	Power

J. Calautit et al. (eds.), *Towards Sustainable Cities in Asia and the Middle East*, Sustainable Civil Infrastructures, DOI 10.1007/978-3-319-61645-2_8

P	Pressure
r	Augmentation Factor
V	Velocity
BL	Boundary Layer
CFD	Computational Fluid Dynamics
CSA	Cross-Sectional Area
DAWT	Diffuser Augmented Wind Turbine
HAWT	Horizontal Axis Wind Turbine
HAWT w. Concentrator	Horizontal Axis Wind Turbine w. Concentrator
WECS	Wind Energy Conversion Systems
WT	Wind Turbine

1 Introduction

Global energy demands are continuously increasing as global population growth accelerates, leading to greater levels of energy consumption and waste with buildings accounting for 40% of the total global energy consumption. This currently, is thirteen years on from the Kyoto Protocol in 1992 (UN 1998) that established an international treaty on the need to address global warming and the effects of climate change. In December 2015, an international pledge in Paris (UN FCCC 2015) was made to fund projects (worth $100 billion) in reducing carbon emissions, controlling the global temperature rise and strengthening the shift toward renewable energy generation.

The current fossil fuel-based economy is still strong, but realistic power generation techniques from renewable energy are emerging, such as power from wind energy, which is one of the most competitive and efficient alternatives to non-renewable power generation. There is an abundance of available wind energy on-shore and off-shore for power conversion. Horizontal Axis Wind Turbines (HAWT) and Vertical Axis Wind Turbines (VAWT) are the most common configurations of which there have been many variations. The three-bladed HAWT has been leading the Wind Energy Conversion Systems (WECS) sector in terms of efficiency, large-scale deployment and cost-effectiveness.

The exergetic performance of a Wind Turbine (WT) can be improved with the use of passive technology such as a Wind Concentrator, a Shroud or more specifically, a Diffuser to form a Diffuser Augmented Wind Turbine (DAWT). This aerodynamic device is able to 'funnel' incident air to the rotor and accelerate its flow such that the rotor can be driven for a higher power extraction threshold over a more controlled, continuous (less intermittent) period of operation. A Diffuser can significantly increase mass flow rate through a WT and maintain low (ideally sub-atmospheric) pressures at exit which is very favourable for accelerating flow (Igra 1981). Currently, power augmentations between 2 and 4 can be achieved, i.e. a power output from a DAWT can be 2–4 times greater than a conventional WT with the same swept area (Ohya and Karasudani 2010). The Diffuser's capability for effectively capturing incoming wind means it can work in low and high intensity turbulent wind conditions and speeds. The DAWT can be used on a small to large scale basis and in the civil environment (building integrated) where wind can be turbulent and hard to capture (Walker 2011).

1.1 Building Integrated Wind Turbines (BWT) with a Perspective into the MENA Region

Over the last few decades, the MENA region has witnessed incredible economic and population growths. This has inevitably meant an increase in building construction in the GCC (Gulf Cooperation Council) where cities are rapidly increasing in size. The GCC members are among the highest energy consuming countries per capita and largest producers of harmful emissions (Asif 2016). The priority now is to re-assess civil infrastructure to improve its Energy Use and Energy Demand (EUED). The Leadership in Energy and Environment Design (LEED) programme is a rating system used to assess Green Buildings across the UAE. In Abu Dhabi, UAE, The Urban Planning Council addressed issues such as "Site Development and Layout" and "Energy and Thermal Efficiency" of buildings in their mandate towards encouraging sustainable practices in building design (pg. 35, Abu Dhabi UPC 2010).

Small-scale WT's are often considered for use in the built environment. In addition to smaller weights and loading, smaller rotor areas allow starting rotation at lower torque in the low and turbulent wind speeds than medium-large. They also respond much better to variation in wind directions and are easier to maintain and replace unlike large scale WT's. In dense urban environments, space is very much sought-after so siting and positioning a WT becomes an art. Figure 1 (left) shows some of the ways a WT can be installed onto a building; side-mounted, roof-mounted, ground-mounted or building integrated (Smith *et al.* 2012). Figure 1 (right) is a real example of the Bahrain Tower which uses building integrated WT's. The structures either side of the HAWT fixtures have been aerodynamically designed to channel wind speed and direction onto the rotors. However, typically WT's installed in built areas are not very popular due to poor performance and high maintenance. There are new techniques and technologies that can address this issue, as will be seen further in this study.

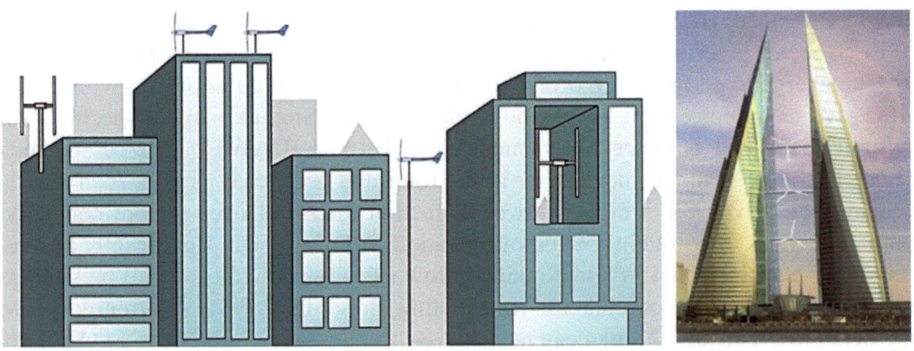

Fig. 1. Left: various siting possibilities for installing a WT (Smith *et al.* 2012), right: the Bahrain tower (Toja-Silva *et al.* 2013)

1.2 Aim and Objective

The aim of this paper has been inspired by the Sustainable Green Building Design Criteria, especially in terms of the Energy and Thermal Efficiency of Buildings. The objective of this paper is to present a CFD study to assess the use of a diffuser and concentrator to enhance the performance of a roof-mounted HAWT. Included is a brief understanding of the theoretical background for a DAWT, the Methodology employed to carry out an analysis, the results and a discussion describing further recommendations.

1.3 Hypothesis

It is expected that installing a Diffuser or Concentrator on a WT will add to the power extraction capability of the WT as air will be accelerated through the rotor resulting in higher velocity and therefore larger output power. Of the two Shrouds applied, it is expected that the diffuser will perform better as it can generated a more negative sub-atmospheric pressure in the wake that will be able to draw in more air mass flow through the rotor thereby increasing power extraction further.

2 Theoretical Background for a Dawt

A DAWT operates in a similar way to a conventional HAWT. The analysis is conducted for subsonic flow. In theory, a diffuser can be employed in low and high wind speeds; this analysis considers low speeds only because at larger Mach numbers, the flow will characteristically change in terms of compressibility. An ideal Diffuser will allow for perfect streamline flow of the working fluid. Flow is symmetrical, there is no separation and frictional losses are negligible. The cross section of the diffuser is an aerofoil as shown in Fig. 2. This generates a lift by the flow through the rotor which in turn generates a velocity that draws in a greater mass flow of air (Hansen 2008).

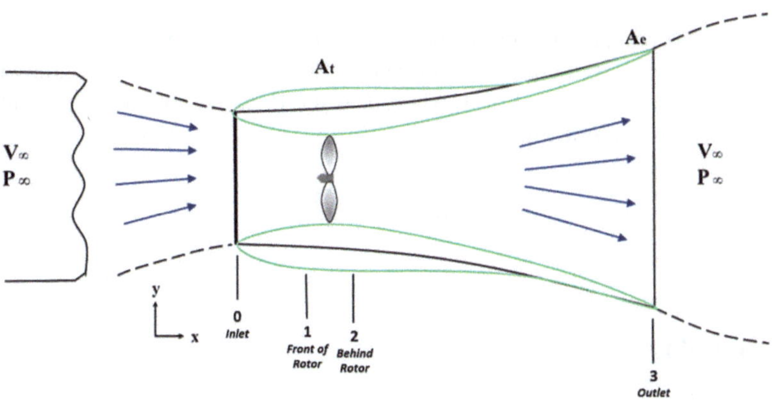

Fig. 2. Schematic of a typical diffuser. The stations 0,1,2 and 3 refer to static locations in the diffuser and the subscripts t, e and ∞ refer to the throat, exit and far upstream/downstream conditions respectively

A number of assumptions have been made. These include the established understanding of the Conservation of Mass (the amount of air entering and leaving the diffuser is the same, no air is lost), Conservation of Energy and Conservation of Momentum where, the rate of change of momentum is equal to all the forces acting in the diffuser in the x-direction (there are no shear forces).

The working fluid is air, so it is important to establish the flow conditions. Air is assumed an ideal fluid, i.e. it is incompressible and inviscid. Air flow is assumed streamlined through the CV, work done due to shear forces is zero, heat exchanges and mass transfer's do not occur. This means the total energy per unit volume of air is constant. Another important consideration is that the entire flow field of the diffuser is one-dimensional and steady. This however, is only valid if flow separation does not occur. Equation 1 shows how to calculate output power, $\boldsymbol{P}$. ρ is the air density (1.225 kg/m^3), A_t is the throat area of the diffuser which is the same as the rotor CSA, V is the air Velocity and a is the axial induction factor (measures the velocity drop from upstream to rotor)

$$\boldsymbol{P} = \frac{1}{2}\rho A_t V 4a(1-a)^2 \tag{1}$$

Assuming an ideal rotor, the axial induction factor is equal to 1/3 and so the Power Coefficient is the maximum limit, i.e. the Betz Limit defined as:

$$C_{p,max} = 4a(1-a)^2 = \frac{16}{27} \sim 59.3\%. \tag{2}$$

The augmentation factor, Eq. 3, is defined as the ratio of power output from a DAWT compared to an ideal (actuator-disk) bare WT of equal blade diameter and operating in the same free-stream conditions (Igra 1981).

$$r = \frac{C_D}{C_{p,max}}\left(\frac{V_t}{V_\infty}\right)^3 \tag{3}$$

To assess and compare the performances of DAWT's another crucial parameter to include in the analysis is the turbine load factor. Turbine load factor (also known as the disk loading coefficient), C_D is Eq. 4, where P is the Pressure computed in front of and behind the rotor as seen in Fig. 2.

$$C_D = \frac{P_1 - P_2}{\frac{1}{2}\rho V_t^2} \tag{4}$$

3 Methodology

CFD is a very useful tool in DAWT analysis as air flow can be virtually assessed without having to use a wind tunnel. A variety of studies on many different parameters can be performed without physical constraints, this saves time and cost and increases repeatability of results. In this study, a 3-bladed roof-mounted HAWT with 3.8 m diameter and height 6.4 m was designed and tested. Three independent studies were conducted to compute and compare the velocity and pressure profiles over the turbine. In all three cases, the turbine was kept constant, in the same location and orientation. A diffuser with NACA 0012 cross section was designed with an area ratio of 0.13 (A_t/A_∞) and length 4.45 m. The diffuser and concentrator are geometrically identical with the rotor at the throat. The concentrator is positioned in the opposite axial direction to the diffuser. The local average wind speed on a daily basis has been recorded using the University's Wind Station as 4 m/s. This has been used to generate the characteristic power curve for the HAWT as seen in Fig. 3.

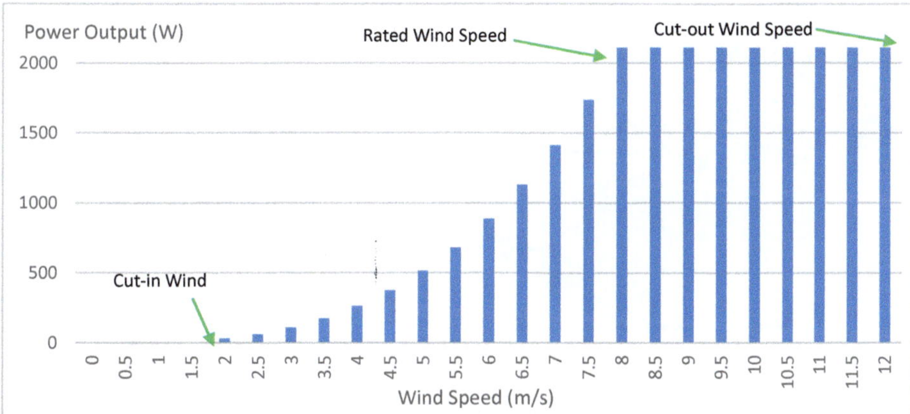

Fig. 3. Characteristic power curve for the design WT

3.1 Computational Domain and Model Definition

Three models were designed in SolidEdge, as shown in Fig. 4. It is important to note that the concentrator converges from inlet and the diffuser diverges from inlet, both follow identical aerofoil profiles.

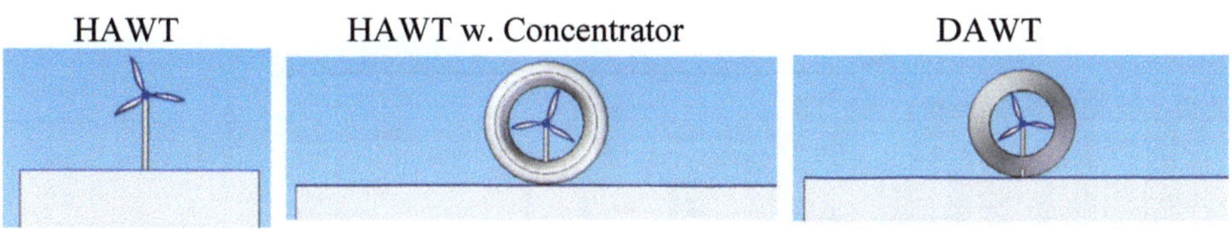

Fig. 4. Left to right: front views of HAWT, HAWT w. Concentrator, DAWT.

The simulations were computed in ANSYS Fluent 14.5 using the RANS k-ε turbulence model in steady state conditions. One million cells were used to mesh the domain; their details are shown in Table 1. The turbine and diffuser/concentrator had curved geometries so unstructured elements was used with negligible distortions as seen in Fig. 5.

Table 1. Domain parameters, the velocity and pressure conditions were kept constant for all cases

	All domains		Boundary conditions	
	Nodes	Elements	Inlet	Outlet
HAWT	146589	792529	Velocity = 4 m/s	Pressure
HAWT w. Concentrator	192780	1032503		
DAWT	192323	1029623		

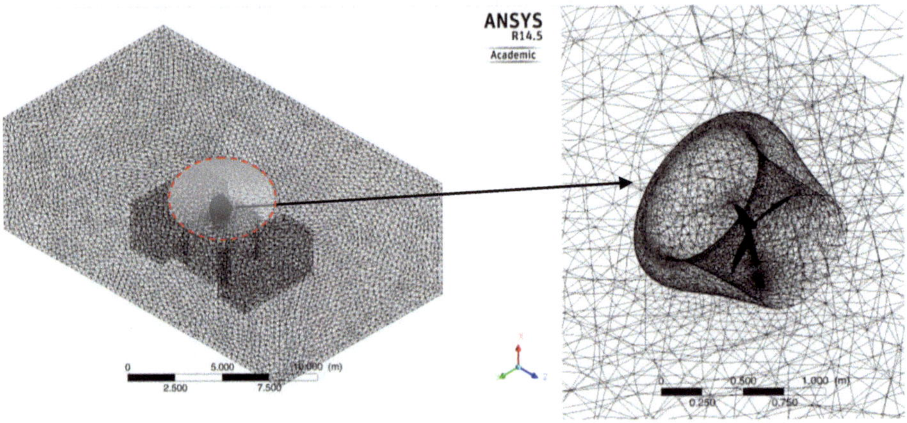

Fig. 5. Left: mesh domain, right: enlarged view of the unstructured mesh used for the diffuser and building

The WT position was placed at the centre of the simulated building as shown in the front view of the SolidEdge model in Fig. 6. Here the incident wind is directly facing the turbine going into the page. The east wing of the campus was simulated as this is where the wind station is installed and where the wind direction is most favourable.

Fig. 6. Left: the Heriot-Watt Dubai campus building, right: as indicated, the section of the building used in the simulation

4 Results

The main parameters that can be reliably used to assess the performance of a WT system are the velocity and the pressure. It is from these results that the equations in Sect. 2 were used to assess the power output and power augmentation for all three cases.

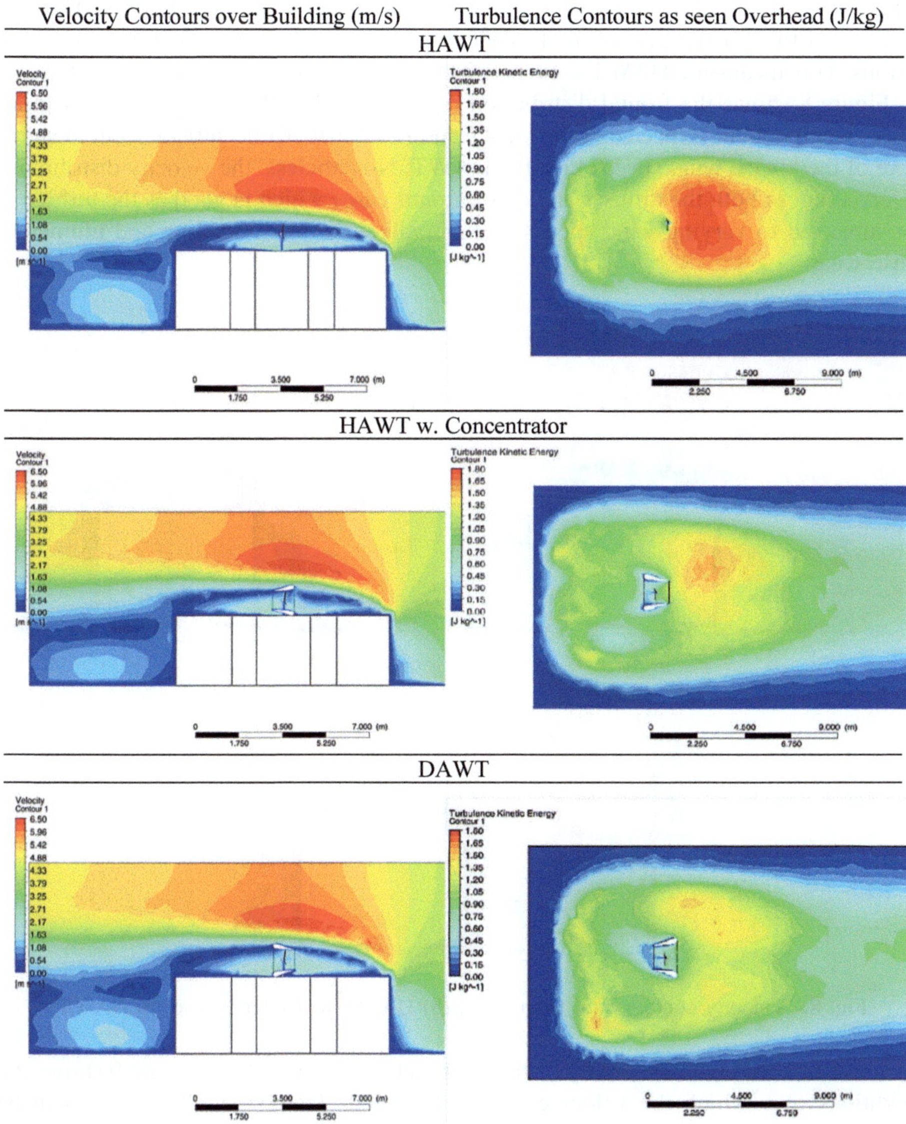

Fig. 7. Velocity and turbulence contours. The wind direction is from right to left in all velocity profiles.

4.1 Velocity Profiles

Progressing down Fig. 7 and it can be seen that the velocity magnitude increases across the rotor. The bare WT is subjected to very low, if not negligible wind speeds. However, with the concentrator and diffuser this significantly improves. It is worth noting that the incident wind onto the building loses much energy as the wind speeds suddenly drop. This can also be seen in the turbulent profiles. In the ideal situation, flow in the wake is desired to be turbulent and show some strong wake rotation, i.e. swirls and Eddy's. There is evidence of this for the case of the DAWT, where there are two distinct wake rotation regions. This means the DAWT is successful at drawing more air through the rotor.

Figure 8 shows the frontal distribution of velocity. The flow is symmetrical in all cases, which is to be expected. The HAWT and HAWT w. Concentrator seem to suffer more of the lowest wind speeds than the DAWT. Nonetheless, the velocity distribution is very small in both the diffuser and concentrator cases which is good as the turbine is not subjected to varying wind speeds that can affect the blade loading and ultimately cause wear.

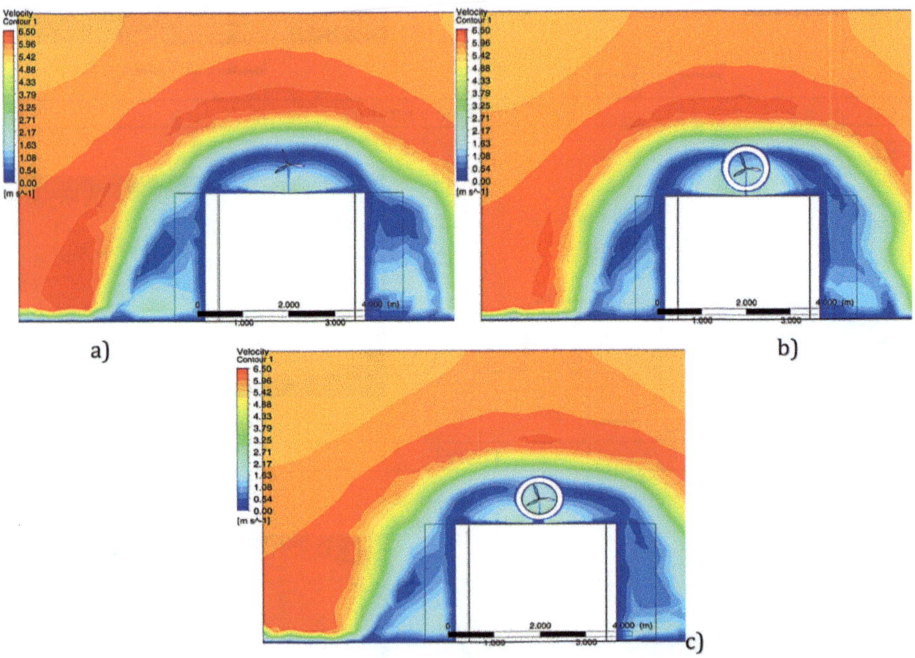

Fig. 8. Velocity contours: (a) front view, (b) HAWT w. Concentrator, (c) DAWT

It can be seen that the velocity distribution radially is important. Figure 9 shows the conditions over which the velocities were computed. The HAWT's velocity can be explained by the uncontrolled conditions it is operating in. It sees a sudden decrease in velocity approaching the centre of the rotor which shows very high tip speed ratios. The augmented turbines however, show more resilience as the variation is not so dramatic because they operate in more controlled conditions.

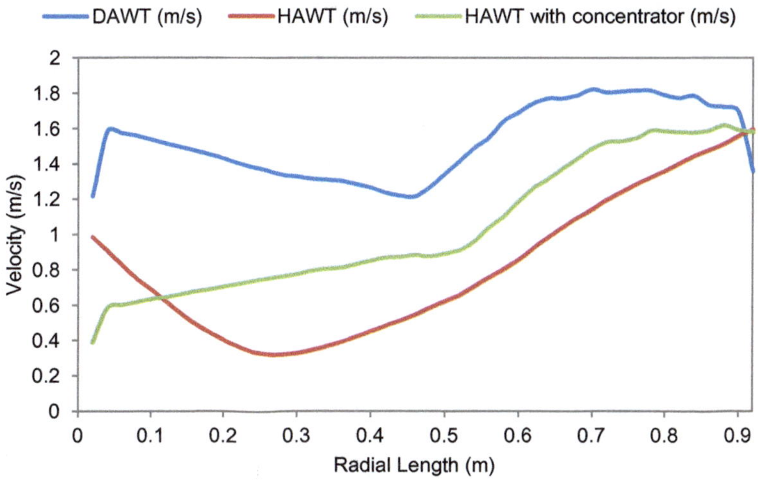

Fig. 9. Variation in velocity profile in the radial direction for the three computed models

Figure 10 shows the domain boundary with length 20 m. The computed lines extend in the axial direction passing through the centre of the DAWT. The rotor was positioned at 9.6 m from right to left in Fig. 10. The axial variation in velocity is more noteworthy. On approach to the building, the profiles are almost identical. After leaving the rotor, the air velocity recovers better in the HAWT than the other two cases which indicates that less energy was extracted by the HAWT. These results agree well with the turbulence results in Fig. 7.

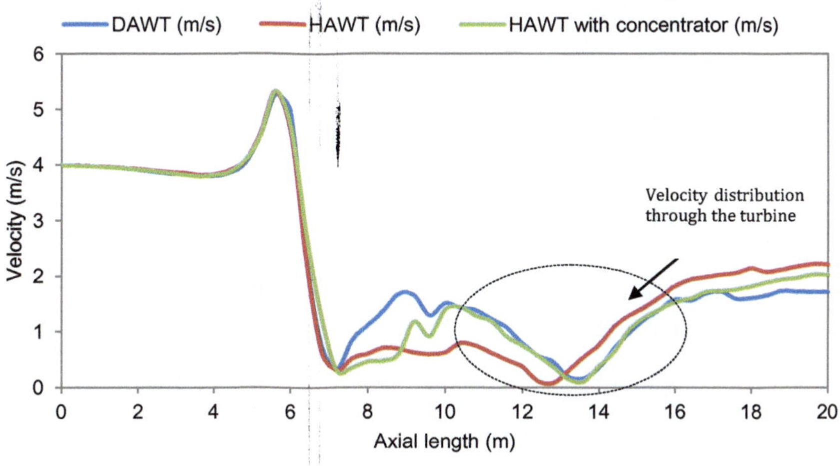

Fig. 10. Variation in velocity profile in the axial direction for the three computed models

4.2 Pressure Profiles

The gauge pressure was computed as shown in Fig. 11. At best performance, the exit pressure should be as low as possible as this generates a large pressure difference across the rotor which encourages a greater mass flow of air (Igra 1981). The DAWT exhibits the lowest exit pressures and the contouring shows the greatest pressure difference for

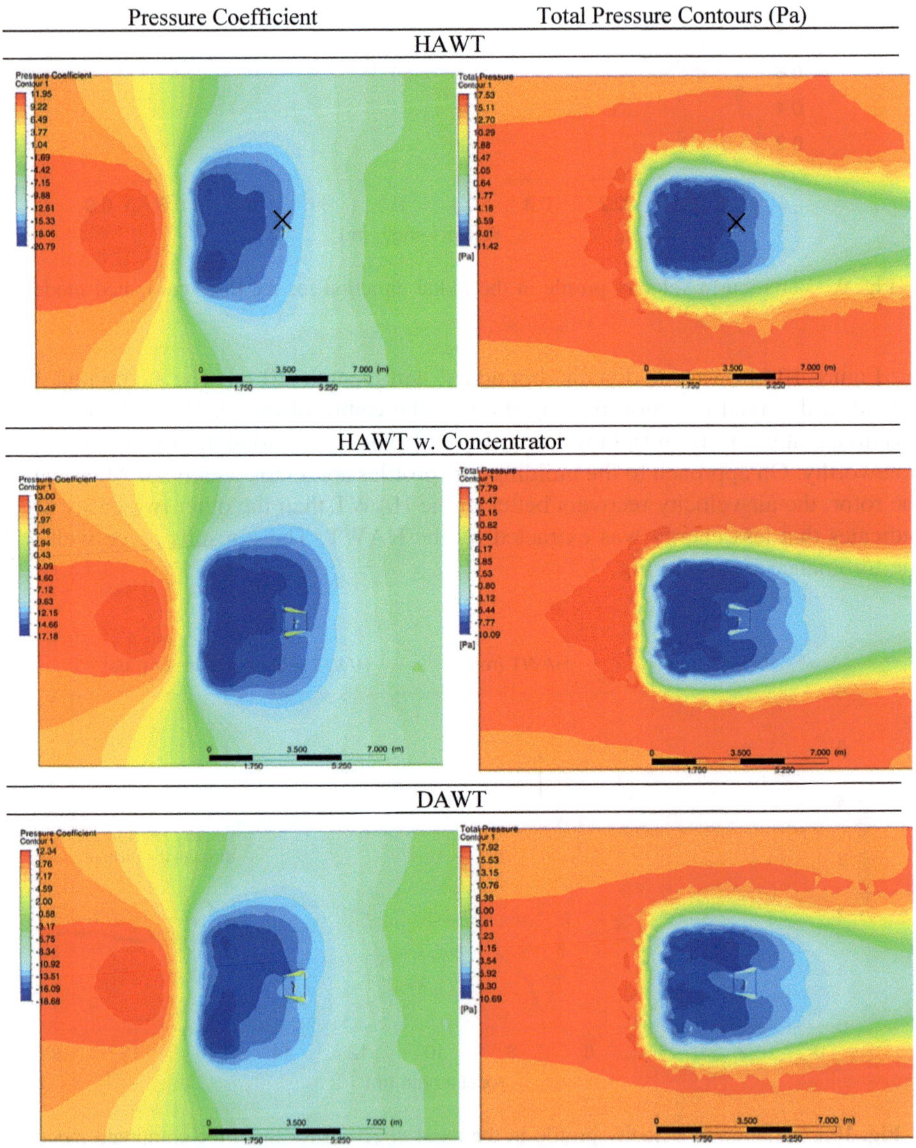

Fig. 11. Pressure coefficient and total pressure contours for all computed cases. The black X on the HAWT results indicates the position of the turbine

this turbine system. The advantage of this can be explained with Eq. 4, where the larger the pressure difference, the larger the turbine disk loading, which in turn will increase the power augmentation (Eq. 3) for a given throat velocity. This has been proven in the next section.

4.3 Performance Assessment

Table 2 and Fig. 12 show the summarised results. The axial positions from 7 m to 14 m were computed exclusively as they cover the region of the WT and diffuser/concentrator. The actual rotor location is at about 9.6 m where the power was calculated. The DAWT outperforms the other systems as it induced the largest pressure drop across the rotor and the largest wind speed. The power augmented for the DAWT was twice as much greater than for the HAWT w. Concentrator, which shows that a diverging shroud from inlet allow for better aerodynamic characteristics.

Table 2. Results table showing the effect of the main parameters on the performance of the different WT systems simulated

	Parameters		Results	
	$V_t[m/s]$	$(P_1 - P_2)[Pa]$	C_D	r
HAWT	0.600141	1.210	5.4849	-
HAWT w. Concentrator	0.924073	1.165	2.2274	1.2072
DAWT	1.299990	2.830	2.2993	**2.4663**

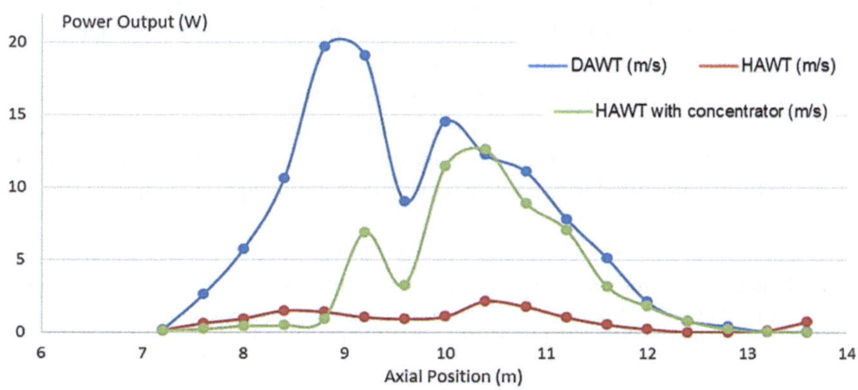

Fig. 12. Variation in power output across the turbine as compared across the three cases

5 Discussion

The results of this study show that there exists potential for roof-mounted DAWT's, since both the wake velocity and exit pressure conditions equate to favourable rotor inlet conditions. The growing potential of DAWT's is very encouraging in the application of the built environment. The possibility of being able to generate more power from a given WT using passive technology such as a diffuser is a key example of innovation and energy efficiency. Experimental testing is required to further this research area.

Building and roof designs can have significant effects on inlet air velocity to a WT as well as the geometric and aerodynamic characteristics of a shroud. Many investigations have been performed to assess the feasibility of wind energy potential in the civil environment. It is crucial to understand the natural wind velocity profiles available for a given location to identify appropriate installation location for any given DAWT (Nishimura *et al.* 2013). The three WT's investigated in this paper were chosen at the centre of the roof to compare wind profiles over equal distances. New methods for siting WT's will be very useful for optimising the civil environment toward the development of smart and sustainable cities (Abohela *et al.* 2013). This could be considered in future studies of DAWT's on roofs. Krishnan and Paraschivoiu (2015) conducted a 3D CFD analysis of a building integrated DAWT using a VAWT as the turbine. They assessed the most appropriate inlet geometry for the DAWT on the roof of a building. They were able to increase the coefficient of power from 0.135 to 0.34 with the use of the shroud. It has been stipulated that compared to conventional WECS, DAWT's are considered safer and quieter due to their smaller size, protective shroud and lower rpm (Kosasih and Tondelli 2012). Bird and bat strikes are anticipated as highly unlikely which is especially favourable considering the implementation of DAWT's into the built environment.

6 Conclusion

It can be concluded that a shroud improves performance, but the use of a diffuser increases performance even further. The maximum power augmentation achieved was 2.5 which agrees very well with literature. Where the HAWT produced negligible power output, the DAWT was still able to produce 20 W at a rotor speed of 1.3 m/s where the local average wind speed is 4 m/s. Further work is required to enhance both the Diffuser aerodynamics and building topology to significantly improve the power out of the DAWT. Nonetheless, it proves a successful conceptual improvement on the conventional bare WT.

Acknowledgments. The research presented has been supported by the funding from Heriot-Watt University, at the School of Energy, Geoscience, Infrastructure and Society.

References

Abohela, I., et al.: Effect of roof shape, wind direction, building height and urban configuration on the energy yield and positioning of roof mounted wind turbines. Renew. Energy **50**, 1106–1118 (2013)

Abu Dhabi Urban Planning Council: Plan Abu Dhabi 2030, Urban Structure Framework Plan, Abu Dhabi Urban Planning Council (UPC), pp. 1–186. Mandate, Abu Dhabi, UAE (2010)

Asif, M.: Growth and sustainability trends in the buildings sector in the GCC region with particular reference to the KSA and UAE. Renew. Sustain. Energy Rev. **55**, 1267–1273 (2016)

United Nations Framework Convention on Climate Change: UN FCCC Paris Dec 2015 Agreement. T.-f. S. Conference of the Parties, Paris, pp. 1–32 (2015)

Hansen, M.O.L.: Aerodynamics of Wind Turbines, pp. 1–175. Earthscan, London (2008)

Igra, O.: Research and development for shrouded wind turbines. Energy Convers. **21**, 13–48 (1981)

Kosasih, B., Tondelli, A.: Experimental study of shrouded micro-wind turbine. Procedia Engineering **49**, 92–98 (2012)

Krishnan, A., Paraschivoiu, M.: 3D analysis of building mounted VAWT with diffuser shaped shroud. Sustain. Cities Soc. **27**, 160–166 (2015)

Nishimura, A., et al.: Wind turbine power output assessment in built environment. Smart Grid Renew. Energy **4**(1), 1–10 (2013)

Smith, J., et al.: Built environment wind turbine roadmap, pp. 1–58. National Renewable Energy Laboratory (NREL), Colorado, USA (2012)

Ohya, Y., Karasudani, T.: A shrouded wind turbine generating high output power with wind-lens technology. Energies **3**(4), 634–649 (2010)

Toja-Silva, F., et al.: Urban wind energy exploitation systems: behaviour under multidirectional flow conditions—opportunities and challenges. Renew. Sustain. Energy Rev. **24**, 364–378 (2013)

UN: Kyoto Protocol to the United Nations Framework Convention on Climate Change, UN, pp. 1–21 (1998)

Walker, S.L.: Building mounted wind turbines and their suitability for the urban scale—a review of methods of estimating urban wind resource. Energy Build. **43**(8), 1852–1862 (2011)

Wang, S.H., Chen, S.H.: Blade number effect for a ducted wind turbine. J. Mech. Sci. Technol. **22**, 1984 (2008). doi:10.1007/s12206-008-0743-8

Safety Through Design: A BIM-Based Framework

Fernanda Rodrigues[1(✉)], Joel Estrada[2], Flávio Antunes[2], and Paul Swuste[3]

[1] RISCO – Department of Civil Engineering, University of Aveiro, Aveiro, Portugal
mfrodrigues@ua.pt

[2] Department of Civil Engineering, University of Aveiro, Aveiro, Portugal
{joeloliveira,flavioantunes}@ua.pt

[3] Delft University of Technology, Delft, Netherlands
P.H.J.J.Swuste@tudelft.nl

Abstract. BIM-based tools have been developed for attain the different Building Information Modeling (BIM) dimensions. However, in spite of the health and safety concerns those tools do not integrate the occupational risk prevention measures. So, in the scope of construction management the production of an integrated framework aiming an overall construction project, including in the 3D model and in the construction planning, the occupational risk prevention and the safety planning is needed. This work presents a BIM-based integrated framework that includes in the BIM model relevant information for risk prevention against falls from height during the construction phase. All this information is used during the construction planning simulation, to optimize production and safety scheduling. It contributes to a design and construction phase with higher safety level, using the potentialities of a BIM model to obtain optimized results integrating in the design elements constructability and safety requirements. The BIM model of a building was developed in detail in Revit-Autodesk, in which structural elements as well safety collective equipment was modelled as all the constructive details. A WBS – Work Breakdown Structure for the building was designed and the construction planning and the construction simulation were done with specific software. This simulation and visualization permit to identify hazards and risks caused by site constraints, construction sequences, and temporary structures, co-existence of different activities, manpower and machinery. In summary, with this integrated BIM-based framework it is possible to incorporate safety since the early stage of construction design and attribute to each detail the safety features and requirements, being all these elements related with construction planning. It is also important to remark the possibility of templates definition that permit the automatic extraction of safety procedures, which can be part of the safety plan and of the safety file of the building.

Keywords: Falls · Height · Prevention through design · 3D model · Planning

J. Calautit et al. (eds.), *Towards Sustainable Cities in Asia and the Middle East*,
Sustainable Civil Infrastructures, DOI 10.1007/978-3-319-61645-2_9

1 Introduction

Occupational health and safety at construction sites arisen with construction activities and is strongly related with the specific activities of this sector involving work at dangerous conditions (at height, in deep levels, in the vicinity, under or over places that include high hazard conditions, work with dangerous equipment, materials and substances, etc.). These different and high dangerous conditions at work lead to label construction as high occupational risk sector as explicit in the Temporary or Mobile Sites Directive (Directive 92/57/EEC). In spite of the requirements of this Directive, planning for safety is still paper based consisting on the identification of potential hazards and of safety measures being generally done separately between design and execution phase and involving different players. The lack of communication and integration between designers and safety coordinators/engineers lead to difficulties to analyse the type, the moment, the location and the reason to implement efficient safety measures [1]. The deficiency of correspondence between the design and the execution phase, and the lack of specificity, synergy and effectiveness between safety planning and the design and construction process are main problems, usually considered separated, that effectively contribute to different levels of construction management failures along with all the construction life cycle. So, design was identified by Directive 92/57/EEC as the key role of optimising safety, which is strictly and continuously linked with constructability since the earlier design phase, and is recognised the crucial importance of the simulation of these aspects during design to avoid failures during the construction and use phase, being Building Information Modeling (BIM) methodology pointed out as a reliable process to achieve that goals [2]. The increasing of Construction design complexity lead to new design and construction methods development, being BIM one of those methodologies. BIM is a process of generation and management of all the information corresponding to the building in the design and execution phase and during all its life cycle. This methodology uses 3D models that contain all the information about geometry, spatial relations, constructive processes, quantities and materials properties [3]. The growing application of BIM in the Architecture, Engineering, Construction and Owners (AECO) sector, as a modeling technology and associated set of processes to produce, communicate, and analyse building models, enables a building to be represented by intelligent objects that carry detailed information about them and also understand their relationship with other objects in the building model. This methodology permits the creation of digital 3D models of buildings with embedded information from design through construction and operation phase. Supported on literature review this work pretends to remark the importance of the development of a BIM-based integrated framework that includes in BIM models relevant information for occupational risk prevention against falls from height during the construction phase. With this goal it was identified a rule-checker BIM-based framework that permits to check the legal and technical requirements contributing to a design and construction phase, with higher safety level. So, it is intended to show the potentialities of BIM when used to obtain optimized results integrated in all the design elements considering constructability and safety requirements.

How BIM 3D models facilitate the integration of health and safety (H&S) practices, through a process of detection and elimination of hazards and risks against fall from height, and how the scheduling based on BIM 4D dimension, permits to integrate safety and constructability since a building design phase, are the main specific research questions of this work. The aim of this paper is to present the framework of a BIM based "Rule-checking" method for the design phase supported on the safety measures established on the legal safety requirements. The "Rule-checking" permits to verify if the established rules in the safety requirements are interpreted and used as a verification of the building model, since the early design phase. Its goal is to be a support for designers, engineers and H&S technicians during the design phase contributing for the reduction of errors and hazards caused by constructive solutions.

2 Occupational Accidents Causality in Construction

Heinrich [5] was identified as who firstly highlighted that the occurrence of industrial accidents is not only a matter of luck but is caused by human unsafe behaviour and unsafe exposure, by Rowlinson and Jia [4], indicating that adequate production management procedures and methods could eliminate risks. This was after developed by Bird [6] in his domino theory. For construction industry, Lingard and Rowlinson [7] developed a behaviour based model, stressing the importance of management infrastructure in influencing individual behaviours'. Reason [8, 9] developed the Swiss Cheese model of accident causation based on an acceptance of the fallible nature of human beings and a tolerance of individual behavioural errors. In this model safety is conceptualised as an outcome of the complementary effect of multiple layers of systemic defences, aiming to eliminate the trajectory of accident opportunity, in which deficiencies are expected in every layer. The emergence of new perspectives responding to the organisations' needs, led to more recent theories as the multiple causation models, systems theories and socio-technical approaches [10]. To answer the specificity of the construction, sector some accidents causality models were developed [11–17], being some summarised in Table 1 [11].

Two of these models were inspired in the Swiss Cheese model: the Loughborough Construction Accident Causation (ConAC) model [18], developed from the study of non-fatal incidents, setting construction accident causation into three levels (immediate causes, determining factors and originating influences [19, 20]; and the systemic model, developed by Hale et al. [21] from the study of fatal construction accidents, that indicate accident causal factors into four levels: the environment, corporate systems, delivery systems, and output from delivery systems. Its application [22–24] proved the importance of looking beyond the immediate/primary circumstances of an accident to identify the failures at the upstream of construction projects such as the design, client actions, education and economic environment in accident investigations [11, 25]. Also Frijters and Swuste [26] recognize the influence that designers, architects and structural engineers have on health and safety of the construction sites considering the design choices the determining factors to achieve safe construction practices and the need to integrate safety in the design objectives.

Table 1. Occupational accident models of the construction industry: overview [11, 12].

Model	Authors	Characteristics
Alternative model of the accident causes	[13]	Systematic analysis of accidents identifying a simplified sequence of organizational and management factors that contribute to the origin of accidents at work. These factors are associated with: - Failures in the safety policy; - Failures in the planning of work; - Failures in construction management; - Human error
Research model of root causes of construction accidents	[14]	Approach centred on the worker's behaviour. This model presents three root causes of the accident origin: - Failure to identify hazard conditions; - Decision to act deliberately after identify the hazard condition; - Deliberated unsafely decision without considering the workplace environmental conditions
Constraint model of construction accidents causation	[15]	Assumes that the participant's inappropriate reaction, against possible constraints that may arise during the design and the execution phase, contributes to the occurrence of occupational accidents. The facts contributing to the rise of accidents: - Construction work inadequate planning; - Construction work inadequate supervision; - Workplace disorder; - Inappropriate construction procedure; - Inappropriate behaviour resulting from the realization of unsafe acts committed by the worker
Hierarchical model of the construction accidents causes	UMIST/ Loughborough University (2003) in [16]	Accidents multiple causes approach. The causes responsible for the occurrence of accidents can occur from design to the execution phase. This model considers that the origin of the accident is caused by immediate failures, including failures arising from the interaction between workers, workplace, equipment and materials, which are in turn influenced by structural factors inherent to the worker, the workplace, equipment and materials.

(*continued*)

Table 1. (*continued*)

Model	Authors	Characteristics
		Also the failures occurred during the design phase associated with the financial conditions, the client requirements, the competence of the design team, the work planning, the Project and the risk management, subsequently influence the health and safety during the execution phase, increasing the probability of accident occurrence
Systemic model of accidents causes	[17]	It primarily focuses on understanding the production system characteristics impact in the genesis of hazardous conditions and in the workers' behaviour. This model considers that the tasks unpredictability is associated with the genesis of unexpected dangerous conditions at work with great potential to cause accidents

3 Construction Safety Management

In the construction sector the main difficulties to implement safety are due to his several specific characteristics such as great capital involved, intensive dynamic of work, labour intensity, immobility and long life cycle of the final product; and the singularity of objects that are developed in a specific institutional, economic and social context. This sector also has great fragmentation, great variety of enterprises with different dimension and specialization and lots of casual work [27, 28]. This sector's specificities contribute to increase the difficulty of preventive measures implementation and consequently to the increase of occupational injuries [28]. According to the Portuguese Authority for the Working Conditions, almost 31% of the fatal occupational accidents in Portugal occur in the construction industry and 14.8% are caused by falls and slips [29]. Construction management deals with different problems like the feasibility and interoperability of all detailed design drawings/documents and specific construction methods/specifications, detailed cost estimating based on an accurate takeoff quantity, constructability and detailed constructed schedule, health and safety risks management, among others. Usually it is faced the lack of specificity of all design pieces, and the inexistence of effectiveness and compatibility between construction schedule, safety planning and the construction process. All the construction management process should begin in the early stage of design phase with an integrated vision, because design influences costs, scheduling, safety, durability, constructability, facility management, and any decision made at the initial stage of a project life cycle has greater influence than those made in later stages or during construction phase. According to [30] the design phase, from its first moment, is the ideal moment to influence construction results. Similarly, according to Furst and Mroszczyk cited by Kamardeen [31] the ideal opportunity to influence construction safety is during the inception, concept design and detailed design phases during which designers can influence safety through their design options. So, the ability to influence safety in construction decreases as the project progresses, for that reason it is necessary to address safety in early stages, implementing Prevention Through Design (PTD) that means "Addressing occupational safety and health needs in the design and redesign processes to prevent or minimize the work-related hazards and risks associated with the construction, manufacture, use, maintenance, and disposal of facilities, materials and equipment" [32]. Directive 92/57/CEE and the Portuguese Law [33] that transposed this Directive establish that risk prevention must be integrated during the early design phase. This integration if reliably done will influence the execution of the construction planning, the quality of the work, and the construction, use, maintenance, repair, refurbishment and demolition performance of the construction projects [34] and specifically the occupational risk prevention as considered by different authors referred in [35]. With the increase of construction design complexity, several specialised designers are needed. This diversification requires an accurate communication, collaboration and coordination among them, especially because the organization of the technical teams is separated in terms of information flows, decision-making and involvement [36]. So, communication plays an important role in all project phases for the diffusion and clear understanding of technical options and safety measures [37]. Communication currently tends to be unidirectional, without discussion between the several project stakeholders, being essential to change this procedure and use new technologies in the design phase for safety issues, based on a common model, integrating all the information, permanently available: thus arises Building Information Modelling (BIM).

4 Safety Through BIM

BIM methodology application is growing in the AECO sector, as a modelling technology and associated set of processes to produce, communicate, and analyse building models. According to [38] BIM is a "Shared digital representation of physical and functional characteristics of any built object (including buildings, bridges, roads, etc.), which forms a reliable basis for decisions". It enables a building to be represented by intelligent objects that carry detailed information about them and also understand their relationship with other objects in the building model. This methodology enables the creation of digital 3D models of buildings with embedded information about a project from design through construction and operation phase. In spite of several BIM-based tools applied for attain all BIM dimensions (3D, 4D, 5D, 6D, 7D) have been developed, in the scope of construction management there are not a concern with the production of an integrated framework aiming an overall construction project, including in the 3D model and in the construction planning, the occupational risk prevention and the safety planning [31]. BIM simulation models can support the analysis of the construction schedule, resources, and management costs. Furthermore, within the model, the entire changing process of the structure and construction site can be recorded and evaluated. This helps the deep analysis of the conflict of construction, and detection of collisions. However, current models do not contain information of the building safety analysis [39].

The main reasons to assist safety management with a BIM based rule-checking system are the risk prevention and the elimination of potential hazards. Another major advantage of rule-checking based on BIM 3D models is time-saving, because manual observation requires huge amounts of time and effort, being therefore more expensive [40]. In general BIM based rule-checking offers detailed and visual reports in a short period of time. As [41] refers "the progress computational tools are much more efficient than humans in applying the rule based checker". Thereby a functional model of analysis and management must contain a structural safety analysis, a schedule conflict management, a resource and cost conflict management and a site conflict management [39]. So, a building 3D and 4D model was developed and a framework of a rule checker according the Portuguese safety rules against falls from height was designed. This aims to enable when a constructive element is intended to be checked in the model to appears the safety rules that have to be accomplished. This will be a support for designers in the scope of safety requirements verification.

5 Methodology

In each country the building regulation features are frequently subjective and complex. So, a model-checker has to be adapted to the regulation of each country. So, firstly has to be developed a global basis and after it must apply the local building codes and regulations. This formalisation of this knowledge (that consists on the interpretation of a set of regulations and of the translation of them into rules that can be process by a computer code) provides significant and adequate data necessary for the development of the Building Regulation-specific object modelling.

The formalisation of building regulations in this context, can be achieved by the following three steps: take the necessary regulations for the study of the building; classify it into declarative (easily read by code) and informative (not read by code) data; decompose the declarative and informative data to extract the semantics. A filter system must be used to determine that the rules are easily interpreted by a computer, e.g., when the rules are applied to an element the answer must be true or false. The second filter extracts the informative clauses, the ones that have subjective information that do not suggest a direct meaning, and contains partially adequate data for computer interpretation rules. The creation of a rule-checking should have the following broad stages: object identification, definition of attributes and enumeration; establishment of semantic relationships and object transformation into classes [42]. The conceptual model of a rule-checking system was developed aiming to manage Portuguese legal rules and technical practices against falls from height, using a 3D BIM model and schedule to create an automatic safety checker system during the design phase of a construction project. An algorithm for the automation of a BIM based rule-checking was also developed that contain nine basic steps according to [1], including: Rule mapping, Rule execution and Rule checking reporting. In the first step the algorithm identifies the hazard and the surrounding environment, after that it applies the safety rules, and finally the results are updated to the BIM model. Other categories can be implemented, such as human input during the rule execution phase, and the quantities, budget and schedule of temporary structures needed.

The object of study considered is a radiotherapy building located in Funchal, Madeira Island, Portugal. The building has a reinforced concrete structure, including foundations, columns, beams and slabs. The underground level walls are also in reinforced concrete, and the first floor have masonry walls and a flat roof.

6 Results

Considering the wide range of works that has to be deal with in the construction sector, according to the type of works, final products and safety rules to be accomplished, this work focus on the construction of the concrete structure of the radiotherapy building referred, and in the protection against falls from eight through the borders and holes of slabs. With the aim of developing a framework in the BIM-based model, integrating elements to identify risks and consequently implement control or prevention measures and make precise risk prevention during the design phase, a formalisation has been made, identifying the Portuguese legal safety rules applied to protection against falls from height that can be extracted and translated into computer code. Two main legal regulations were analysed, the former Safety Regulation for Construction Work [43] and the legal rules for working sites in construction [44]. From this one it was concluded that within the 92 rules of the legal document only 12 are applied to the goal of this work. From [43] within the 234 rules 97 are applied to protection against falls. After this each rule was classified into declarative (easily read by code), informative (not read by code) data, and others (not adequate for automatic check). A framework was developed with the primary steps to create a rule-checker system aiming the automatic checks of technical and legal requirements contributing to a high safety level in the design, construction and operation phase, as explained in the previous section.

The building (structure phase) under study was modelled with Autodesk Revit 2015 (Fig. 1), in which were inserted all the required collective protective equipment (temporary structures) to achieve an acceptable level of safety, regarding the prevention against falls from height.

The 3D safety objects and the safety project were developed (see Fig. 2). All this information was considered to be used during the simulation of the construction planning through Autodesk Naviswork 2015 (to obtain the 4D model), in order to

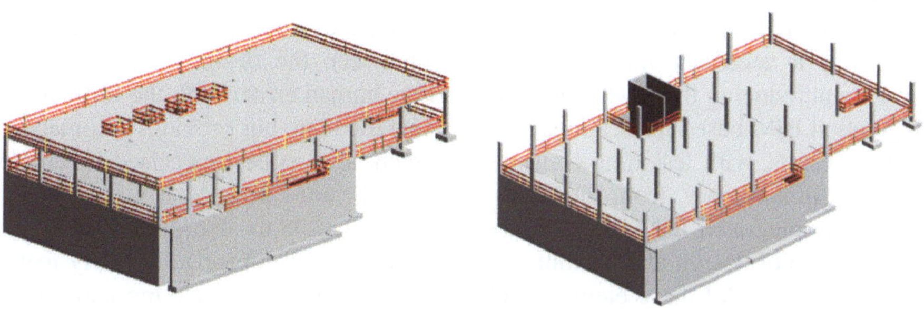

Fig. 1. Building 3D model with the safety protective equipment

3D object model – safety guards | 3D object model – safety line vertical support | 3D object model – horizontal holes protection

Fig. 2. Safety components 3D models.

optimize the construction and safety planning. It was possible to create three-dimensional objects (guard rails, safety cables, horizontal openings (holes) protection), a safety line with its vertical support and it was possible to verify in the 3D model the exact placement of this temporary elements, its temporal sequence (the planning), extract the bill of quantities and costs. Through Revit API's (Application Programming Interface) it was possible to develop third-party plugins that can analyse and interact with BIM models developed using the Revit *software*. Creating such tool allows project managers to easily check the model with an interface that lists the openings on the slabs and confirm the presence of the fall hazards, knowing the places where the safety objects like the horizontal openings protections are needed in an automatic way, consuming less time and reducing the probability of errors. Also, since BIM models are rich in data, it was possible to retrieve more information about the selected openings, and export all the information needed to improve the safety sheets. With the construction planning with Naviswork software it was possible to simulate the construction and the sequence of the temporary safety protective equipment installation.

7 Conclusions

Construction industry still remains one of the most dangerous industries, where more fatal accidents occur, being Prevention-through-Design one of the most effective procedure to combat the hazards at the source. With the development of BIM methodology, it is possible to integrate and develop automatic safety checks (rule-checking), developed through computational methodologies created by an interpretation of the legal and technical regulations, fundamental for PTD. Formalisation of legislation in effect is the first and main process for the creation of a BIM based rule-checking. Although most of the regulations are not easily translated into computer language the majority can be covered by an automated model checker. A BIM based model adds a necessary non-existent bridge between the design phase and safety management reducing the time consuming and the human error. With the formalization of regulation BIM based model does not eliminate the man, but provides automatic and objective results, with less time consuming, which have to be supervised by an H&S technician. In summary, from the research and work until now done it is considered that with an integrated BIM-based framework it is possible to incorporate safety since the early stage of design and attribute to each design detail (object) the safety features and requirements. All these elements related with construction planning and with the activities that have to be performed during the use phase of the building will effectively implement safety through all the building life-cycle. It is important to remark the possibility of templates definition that permit the automatic extraction of safety technical procedures, which can be part of the safety plan and of the safety file of the building. This work is in progress to develop the BIM based check-ruler framework based on a 3D and 4D BIM model complemented with a job hazard analyses that basically will consist on a support for designers' options. The software application to achieve this objective is under development.

References

1. Zhang, S., Teizer, J., Lee, J.K., Eastman, C.M., Venugopal, M.: Building information modeling (BIM) and safety: automatic safety checking of construction models and schedules. Autom. Constr. **29**, 183–195 (2013)
2. Capone, P., Getuli, V., Giusti, T.: Constructability and safety performance based design: a design and assessment tool for the building process. In: 31st International Symposium on Automation and Robotics in Construction and Mining, pp. 313–320, 9–11 July 2014
3. Eastman, C., Teicholz, P., Sacks, R., Liston, K.: BIM Handbook: A Guide To Building Information Modeling for Owners, Managers, Designers, Engineers and Contractors, 2nd edn. Wiley, Hoboken (2011)
4. Rowlinson, S., Jia, Y.A.: Construction accident causality: an institutional analysis of heat illness incidents on site. Saf. Sci. **78**, 179–189 (2015)
5. Heinrich, H.W.: Industrial Accident Prevention, 2nd edn. McGraw, New York (1941)
6. Bird, F.E.: Management Guide to Loss Control. Loss Control Publications, Mount Waverley (1974)
7. Lingard, H., Rowlinson, S.: Behaviour-based safety management in Hong Kong's construction industry: the results of a field study. Constr. Manag. Econ. **16**, 481–488 (1998)
8. Reason, J.: Managing the Risk of Organizational Accidents. Ashgate, Surrey (1997)
9. Reason, J.: Human error: models and management. BMJ **320**, 768–770 (2000)
10. Manu, P.A., Ankrah, N.A., Proverbs, D.G., Suresh, S.: Investigating the multi-causal and complex nature of the accident causal influence of construction project features. Accid. Anal. Prev. **48**, 126–133 (2012)
11. Rodrigues, F., Coutinho, C., Cardoso, C.: Correlation of causal factors that influence construction safety performance: a model. Work - J. Prev. Assess. Rehabil. **51**(4), 721–730 (2015)
12. Azevedo, R.: Acidentes em operações de movimentação manual de cargas na construção. Tese de Dout., Universidade do Minho, Portugal (2010). (in Portuguese)
13. Whittington, C., Livingston, A., Lucas, D.A.: Research into Management, Organisational and Human Factors in the Construction Industry. HSE Research Report No. 45, Norwich (1992)
14. Abdelhamid, T.S., Everett, J.G.: Identifying root causes of construction accidents. J. Constr. Eng. Manag. **126**(1), 52 (2000)
15. Suraji, A., Duff, A.R., Peckitt, S.J.: Development of causal model of construction accident causation. J. Constr. Eng. Manag. **127**(4), 337–344 (2001)
16. Haslam, R.A., Hide, S.A., Gibb, A.G.F., Gyi, D.E., Pavitt, T., Atkinson, S., Duff, A.R.: Contributing factors in construction accidents. Appl. Ergon. **36**(4), 401–415 (2005)
17. Mitropoulos, P., Abdelhamid, S.T., Howell, H.A.: Systems model of construction accident causation. J. Constr. Eng. Manag. **131**(7), 816–825 (2005)

18. Gibb, A.G.F., Haslam, R., Gyi, D.E., Hide, S., Duff, R.: What causes accidents? Inst. Civ. Eng. **159**(6), 46–50 (2006)
19. Hide, S., Atkinson, S., Pavitt, T.C., Haslam, R., Gibb, A.G.F., Gyi, D.E.: Causal factors in construction accidents. HSE research report 156 (2003)
20. Hide, S., Atkinson, S., Pavitt, T.C., Haslam, R., Gibb, A.G.F., Gyi, D.E., Duff, R.A.: Contributing factors in construction accidents. Appl. Ergon. **36**, 401–415 (2005)
21. Hale, A., Walker, D., Walters, N., Bolt, H.: Developing the understanding of underlying causes of construction fatal accidents. Saf. Sci. **50**, 2020–2027 (2012)
22. Cooke, T., Lingard, H.: A retrospective analysis of work-related deaths in the Australian construction industry. In: ARCOM 27th Annual Conference, Reading, UK (2011)
23. Lingard, H., Cooke, T., Gharaie, E.: A case study analysis of fatal incidents involving excavators in the Australian construction industry. Eng. Constr. Archit. Manag. **20**(5), 488–504 (2013)
24. Behm, M., Schneller, A.: Application of the Loughborough accident causation model: a framework for organizational learning. Constr. Manag. Econ. **31**(6), 580–595 (2013)
25. Gibb, A.G.F., Lingard, H., Behm, M., Cooke, T.: Construction accident causality: learning from different countries and differing consequences. Constr. Manag. Econ. **32**(5), 446–459 (2014)
26. Swuste, P., Frijters, A.: Safety assessment in design and preparation phase. Saf. Sci. **46**, 272–281 (2008)
27. Vrijhoef, R., Koskela, L.: A critical review of construction as a project-based industry: identifying paths towards a project-independent approach to construction. CIB Combining Forces, June 2005
28. Swuste, P., Frijters, A., Guldenmund, F.: Is it possible to influence safety in the building sector? A literature review extending from 1980 until the present. Saf. Sci. **50**(5), 1333–1343 (2012)
29. ACT: Acidentes de Trabalho Mortais. http://www.act.gov.pt/(pt-PT)/CentroInformacao/Estatistica/Paginas/AcidentesdeTrabalhoMortais.aspx. Accessed 21 May 2015
30. Potts, K.: Construction Cost Management. Learning from Case Studies, 1st edn. Taylor & Francis Group, New York (2008)
31. Kamardeen, I.: 8D BIM modelling tool for accident prevention through design. In: 26th Annual ARCOM Conference, pp. 281–289, September 2010
32. Manuele, F.A.: Prevention through design: addressing occupational risks in the design and redesign processes. ByDesign, pp. 1–13 (2007)
33. DL n.º 273. Condições de Segurança e de Saúde no Trabalho em Estaleiros Temporários ou Móveis, 29 Outubro, 2003, DR Série A, 251, pp. 7199–7211 (2003). (in Portuguese)
34. Rodrigues, M.F., Maranhão, S.H.J.: A Formação dos Coordenadores de Segurança e Saúde do Sector da Construção. SHO2007 - Colóquio Internacional sobre Segurança e Higiene Ocupacionais, pp. 69–72 (2007). (in Portuguese)
35. Zhou, W., Whyte, J., Sacks, R.: Construction safety and digital design: a review. Autom. Constr. **22**, 102–111 (2012)
36. Dossick, C.S., Neff, G.: Organizational divisions in BIM-enabled commercial construction. J. Constr. Eng. Manag. **136**(4), 459–467 (2010)
37. Ganah, A., John, G.A.: Integrating building information modeling and health and safety for onsite construction. Saf. Health Work **6**(1), 39–45 (2014)
38. ISO 29481-1: Building Information Modeling—Information Delivery Manual—Part 1: Methodology and Format (2010)
39. Hu, Z., Zhang, J.: BIM- and 4D-based integrated solution of analysis and management for conflicts and structural safety problems during construction: 1. Principles and methodologies. Autom. Constr. **20**(2), 155–166 (2011)

40. Zhang, S., Lee, J.K., Venugopal, M., Teizer, J., Eastman, C.: Integrating BIM and safety: an automated rule-based checking system for safety planning and simulation. In: Proceedings of CIB W099 Conference, pp. 1–13 (2011)
41. Melzner, J., Zhang, S., Teizer, J., Bargstädt, H.-J.: A case study on automated safety compliance checking to assist fall protection design and planning in building information models. Constr. Manag. Econ. **31**(6), 1–14 (2013)
42. Malsane, S., Matthews, J., Lockley, S., Love, P.E.D., Greenwood, D.: Development of an object model for automated compliance checking. Autom. Constr. **49**, 51–58 (2015)
43. Decreto-lei 41821. Normas de segurança no trabalho da construção civil, 11 de Agosto de 1958: Min.Obras Públicas e das Corpor. e Prev. Social (1958). (in Portuguese)
44. Portaria n.º 101. Prescrições mínimas de segurança e de saúde nos locais e postos de trabalho dos estaleiros temporários ou móveis, 3 de Abril de 1996, Diário da República, 1ª Série-B n.º 80, pp. 703–706 (1996). (in Portuguese)

Toward the Integration of BIM Energy Saving Concepts

Abdelrahman Hadid(✉), Nabil H. ElAshkar, Wael Kamel, and Hesham Basiouni

College of Engineering and Technology, AASTMT, Giza, Egypt

Abstract. Nevertheless to say that Energy consumption has a negative significant impact on the environment. In this context, the buildings sector accounts for around a percentage of 30% of the world energy total consumption. Consequently, building designers should consider energy savings strategies in their designs to achieve a low level of energy consumption while maintaining their design goals. On the other hand, building energy performance assessments face difficult challenges. The main target of such paper is to evolve a newly easy-to-use integrated BIM-based framework to support decision making at the conceptual stage, which is considered the most critical phase of the design. This framework empowers designers to investigate more design options while achieving the high level of the energy performance. Utilizing the power of visual programming to set the key-parameters that have significant effects on the performance like building orientation and form, and with using the appropriate BIM tools to easily transfer the design from the conceptual phase to the schematic phase of the project. A case study was applied to showcase the implementation of this workflow. Designers with limited programming experience may use this workflow to perform a broad variety of design options analysis, take the conscious design decision and easily implement it with their BIM-based workflow which will effectively reduce the time of design, improve the collaboration between teams, cause a significant energy saving in the building, and inevitably reduce life-cycle cost of the building.

1 Introduction

The global environmental concerns have put an eye to energy consumption and the level of carbon dioxide emissions. The sector of buildings is counted as the biggest world energy consumer and represented 40% of the emissions of carbon dioxide in 2010 and about 30% of the world energy total consumption (US Dept. of Energy 2012). Meantime, the building sector has the best potential for carbon decrease with the most reduced cost (Sustainable Buildings and Climate Initiative 2009). The requirement for sustainable buildings with least environmental effects is increasing. Applying individual sustainable procedures would help to enhance the performance of the building to some extent. However, achievement of high-level energy performance needs applying of the ideal combination of several methodologies (Stevanović 2013). In the meanwhile, enhancing building performance by applying separate methodologies can be simple, but streamlining the design performance faces complicated

J. Calautit et al. (eds.), *Towards Sustainable Cities in Asia and the Middle East*,
Sustainable Civil Infrastructures, DOI 10.1007/978-3-319-61645-2_10

challenges (Wang et al. 2005). The complexity results from the great number of multi-disciplinary interconnected parameters involved in optimizing building performance in addition to the complicated natures of building simulation results (Nguyen et al. 2014). It is a common approach that energy performance analysis comes as a second step after the step of architectural/engineering design. As Schlueter and Thessling (2009) believe that this practice does not count the significance of relating the design and energy analysis processes during early phases. Moreover, it considered a wasteful method of backtracking of adjusting the design in order to accomplish a lot of performance measures.

The good planning project and the inclusion of environmental criteria in the conceptual stage have a great effect on the reduction of energy consumption. Decisions taken at the early stage of the process of design have a significant impact on the result compared to the decisions at the end of the design process, in spite of the fact that the early decisions are based on less knowledge about the goals to be achieved. The core indicators of conceptual design aim to predict the building's sustainability performance at the conceptual stage, where the data available is rare (Bragança et al. 2014).

In an environmental design approach, the initial step is to identify basic and critical natural environmental elements that impact the design of the building. To enhance the possibility of making appropriate environmental design decisions in the planning stage, a conscious understanding of the design aspects and site environmental condition are required even before the evaluation of existence of building form and design. In this phase, designers evaluate their decisions, not according to the result of any analysis, but on according to personal judgment that stems from experience and the understanding of environmental data. An and Zofchak (2012) explained that neglecting basic analysis may drive designers to unnecessary analysis in order to answer basic questions. Weytjens et al. (2012) have discussed the benefits of integrating environmental analysis and building simulation into the design process.

An integrated BIM system can improve collaboration and communication processes among project team in the conceptual phase to submit adequately a well-performing building within the operations. BIM submit relevant building information which is required for building energy performance analysis and if utilized effectively lead to saving more effort and time in preparing input data for building performance simulation while reducing errors. BIM can assist in many sustainable design aspects: building orientation, building massing (to optimize the envelope of the building), energy modeling (reducing energy needs), daylighting analysis, and sustainable materials (Azhar et al. 2009). In recent years, many papers have been published discussing the optimization of building performance but a very limited number of papers have discussed the BIM-based performance optimization (Yan and Rahmani 2015).

Various previous studies have been done to address the important need for parametric design tools integrated with analysis of building performance and to enhance the building design process, (Jakubiec and Reinhart 2011; Lagios et al. 2010; Niemasz et al. 2011; Paoletti et al. 2011; Roudsari and Pak 2013; Shi and Yang 2013). These studies tried to enable designers to explore various design options and access energy performance analysis results quickly. These studies aimed to develop a framework to make the parametric performance-based building design more accessible at the conceptual stage. However parametric BIM-based performance optimization leads to high-performance building design, a very limited number of researches have discussed it. In this paper, an Environmentally-based Programming BIM Use process will be introduced as a framework which overcomes the limitations of the previous research studies.

2 Literature Review

The most of the studies that considered parametric design to achieve high-performance buildings are made by Rhinoceros® (Rhino), a 3D NURBS modeling program, in light of its integrated visual programming tool, Grasshopper®, which supports parametric design (McNeel 2015). Such papers attempt to enhance workflows to integrate Rhino/Grasshopper with building performance analysis tools. For example, Lagios et al. (2010) made a workflow by Rhino/Grasshopper to export scenes (geometry, properties, material, and sensor grids) to Radiance/ DAYSIM to calculate a daylighting series performance indicators. Jakubiec and Reinhart (2011) discussed a design workflow for the analysis of integrating daylighting utilizing Radiance/DAYSIM and thermal analysis using EnergyPlus within their plugin called DIVA. Roudsari and Pak (2013) developed an open and free source plugin which connects Grasshopper to EnergyPlus, Radiance, DAYSIM and OpenStudio for building energy and daylighting analysis. In addition to that, they created Ladybug plugin in order to import weather files of EnergyPlus (EPW) for the best understanding about the data of weather and to visualize the analysis results of building performance. Few studies added the best practice of current tools of the parametric building energy simulation process to assist designers to find an ideal set of solutions for a particular project. For example, Shi and Yang (2013) created a performance driven workflow within integrating Ecotect with Rhino/Grasshopper to acquire an ideal roof shape utilizing a single objective optimization algorithm. Ali and Nassar (2013) utilized DIVA alongside Galapagos, an Evolutionary Algorithm plugin in Grasshopper, in order to optimize the performance of urban daylight.

In spite of the fact that these studies made parametric architectural design workflows with performance optimization available in Rhino/Grasshopper, they don't have entry to the actual available assemblies and properties of building in BIM. In the latest years, many studies have been conducted on building performance optimization, although few of such studies are a parametric BIM-based performance optimization. Yan and Rahmani (2015) presented an integrated framework for the Performance Optimization based on BIM (BPOpt). Such framework empowers designers to investigate design choices utilizing an open source, user interface visual programming tool, Dynamo (2015) on the most widely used platform for BIM, Autodesk REVIT. This framework helped to create various models of building design, evaluate the energy performance of models within the cloud-based simulation, and attempt to find the best alternatives of proper design. Other packages of simulation can be added easily into the framework of BPOpt, such as the package of structural analysis which was used to optimize the structural design (Vermeulen 2015). In addition, Subhajit et al. (2016) has provided a BIM-based Integrated Spatial-Structural Optimization in the Conceptual Design Stage of Project using Dynamo, Space Planning package, and Structural Analysis package. Chong et al. (2016) provided a mixed study to define the current state-of-the-art BIM enhancement for sustainability. Through 91 Academic publications from 2011 to 2016, the planning stage has the least number of publications (just ten papers). Moreover, there is a significant need to provide a structured process to integrate all the above mentioned together in order to boost their benefits.

Process is considered an effective factor in delivering a high-performance green facility within budget and on time, as indicated by Horman et al. (2004) Consequently, the sustainable objectives face additional challenges in the delivery process to achieve the project environmental goals which require an early involvement of key project stakeholders (Beheiry et al. 2006). Building process models define the basic functions required to provide a facility to the end user and assure significant relationships, and information regarding delivering high-performance facility. In return, a building process model forms the basis for developing significant understanding about the characteristics of high-performance delivery of the building.

Like to green building, BIM implementation favors more collaborative project delivery methods and is asking a paradigm change in current processes of business construction (Mihindu et al. 2008). Laiserin (2007) focus on potential fields of enhancing BIM implementation to look at the integration, consistency, accuracy, coordination, and synchronization of the process of BIM implementation. Process modeling that endeavors to address BIM adoption challenges and implementation at the project level, were uncommon before the presentation of Project BIM Execution Planning Guide (Coates 2010). BIM Execution Plan submitted the first well-formatted process model for BIM implementation at the project level. It includes a four-step procedure to enhance a BIM execution Plan. The procedure is designed to steer owners, program managers, and participants in early project within a structured process to enhance detailed and consistent plans for projects. This procedure was developed within a multi-step research process which included industry interviews with more than 40 experts of the industry, detailed analysis of existing planning documents, focus group meetings with industry participants, process mapping research to design an efficient and effective mapping structure, and case study research to validate the procedure (Computer Integrated Construction Research Program 2010). Though, BIM Execution Plan is not mainly designed for targeting green building projects but more about providing a generic process model for any projects that implement BIM.

3 Research Methodology

In order to achieve the previously mentioned objectives, "A conceptual framework for Integrated BIM-based Energy Saving Optimization in Buildings", the steps shown in Fig. 1 were adopted (see Fig. 1).

Fig. 1. Illustration of the methodology procedure adopted in that research

3.1 Illustration of the Current Workflow

BIM Project Execution Planning Guide is considered a well-documented structured process which aims to ensure that all project parties are fully aware of the BIM opportunities associated with the proper BIM implementation at both project-level and organizational-level. It contains twenty-five common BIM Uses providing a good representation of the current uses of BIM in the AEC industry (e.g., design authoring, design coordination, asset management, and record modeling). “BIM Use is a unique task or procedure on a project which can benefit from the integration of BIM into that process” (Computer Integrated Construction Research Program 2010). The team identifies the project goals then chooses the appropriate BIM Uses. A high-level process map will be designed to show the sequencing and interaction between the selected BIM Uses (see Fig. 2). Thus, all team members could understand clearly how their work processes interact with each other. Then, a detailed process map for each BIM Use will be performed by an organization or several organizations such the case for the proposed “Environmentally-based Programming BIM Use” that will be discussed in more details.

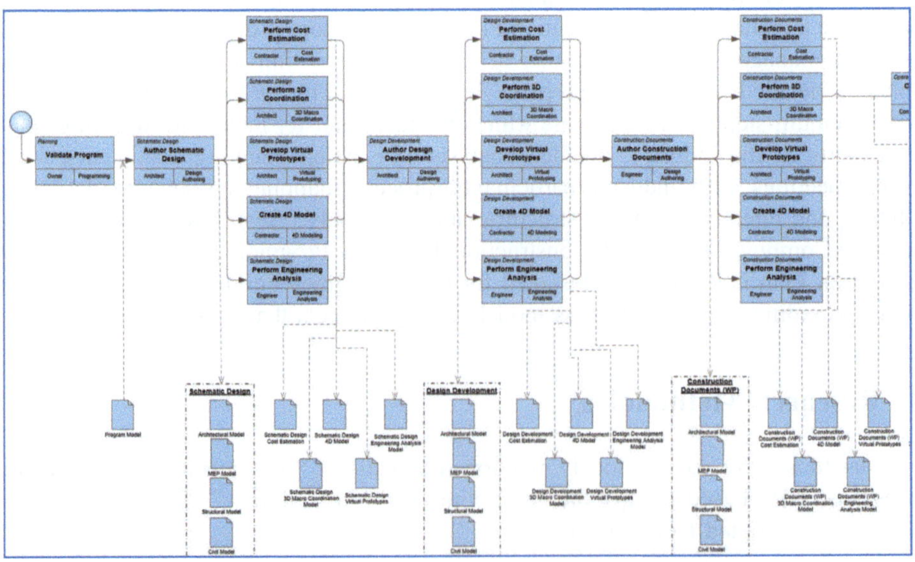

Fig. 2. Portion of Level 1, high level process map. Source: (Computer Integrated Construction Research Program 2010)

3.2 Analysis of the Current Workflow

As mentioned before, BIM Execution Plan is more about providing a general process such as Programming BIM Use process (see Fig. 3). It does not take into consideration the environmental effect of the design at that planning stage which will lead to energy savings in buildings.

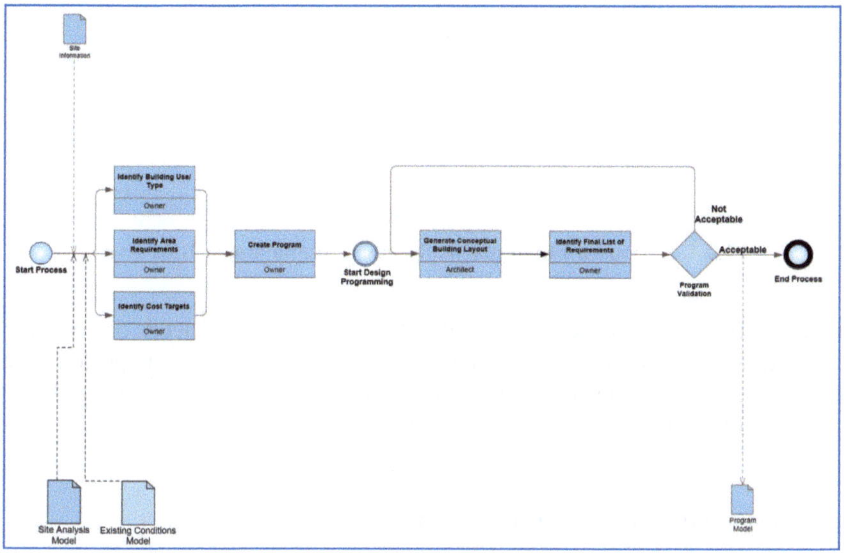

Fig. 3. Programming BIM Use. Source: (Computer Integrated Construction Research Program 2010)

3.3 Proposed Workflow

In this proposed workflow, an implementation of the environmental aspects has been adopted into Programming BIM Use to be "**Environmentally-based Programming BIM Use**" (see Fig. 4).

The first step, the owner identifies Building use/type and Area requirements. The site information and the site model are required data for this first step. In the second step, the owner creates the Program document which summarizes the project needs for area requirements. In the third step, the architect will generate a conceptual building form, which will be ready for environmental analysis which will lead to determining the optimum building orientation for that form. The fourth step is a decision-making step. If the building form and orientation are acceptable for the owner and satisfy the program requirement, then the move to the fifth step will be inevitable. If the answer of the decision was NO, the probable decision whether the program needs to be changed or updated. In the case of changing or updating, the team should move back to the second step in order to change the program document. If the program does not need to be updated or changed, then the third step will be required again to generate another

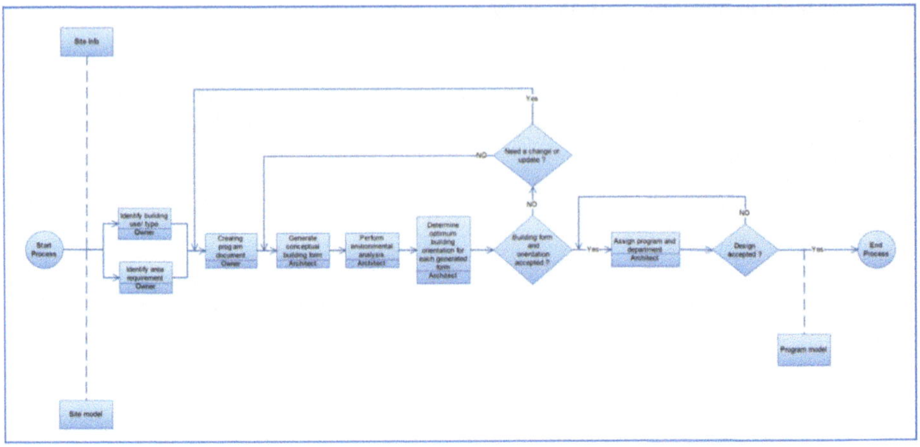

Fig. 4. Proposed Environmentally-based Programming BIM Use

conceptual building form to see whether it will satisfy the owner needs or not. The fifth step is to assign departments, programs, and circulation for the suggested form, if accepted, it means that the design team has successfully completed the process and results in the Program model as a deliverable of that planning stage.

3.4 Implementation of the Proposed BIM Use

The BIM Use is implemented via one of the most famous BIM platforms Autodesk Revit, its visual programming tool Dynamo and its packages; Space Layout, Ladybug, and its optimization tool Optimo (project URL at Rahmani et al. 2015). The BIM Use proceeds as the following steps (see Fig. 5).

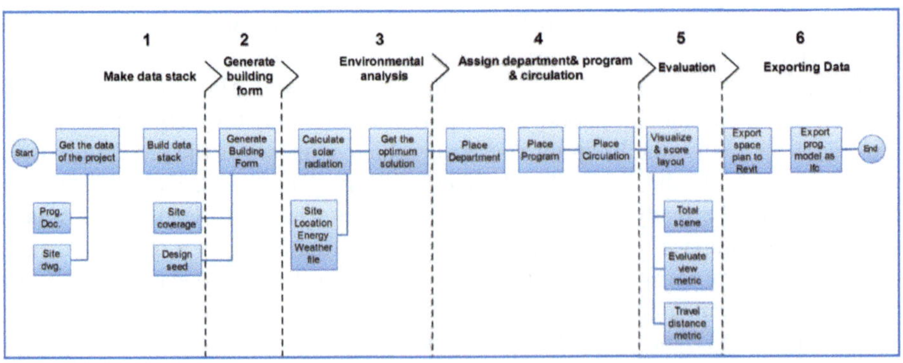

Fig. 5. The proposed process of the implementation of the proposed BIM Use

3.4.1 Make Data Stack

The Dynamo definition needs input information in the form of ".csv" file for program document and ".sat" file for the site layout. That ".csv" file should include information about the required program that will fit into the site as the next order; Program ID, Program name, department, quantity, area, total area, importance value, program type, adjacent programs, department ID, and adjacent departments (see Fig. 6). The Data Stack, which will be retrieved from the input information, is an essential for all later SpaceLayout computation or analysis, as it stores and sorts both departments and program data based on the importance value.

	A	B	C	D	E	F	G	H	I	J	K	L
1	SPACE ID	SPACE NAME	DEPARTMENT	QUANTITY	AREA	TOTAL AREA	PREFERENCE	PROGRAM TYPE	ADJACENT SPACES	DEPARTMENT	ADJACENT DEPARTMENTS	
2	0	Patient Visitor Lounge Waiting Area	PSYCHIATRIC INPATIENT	10	20	200	6	REG	4	0	1.3.4	
3	1	Alcove Telephone	PSYCHIATRIC INPATIENT	1	10	10	10	REG	0	0		
4	2	Alcove Vending Machine	PSYCHIATRIC INPATIENT	1	20	20	10	REG	0	0		
5	3	Toilet Male Female	PSYCHIATRIC INPATIENT	1	60	60	5	REG	0	0		
6	4	Workstation Reception	PSYCHIATRIC INPATIENT	1	40	40	2	REG	0.6	0		
7	5	Lockers Visitors	PSYCHIATRIC INPATIENT	10	0	0	7	REG	4	0		
8	6	Entrance Vestibule	PSYCHIATRIC INPATIENT	1	100	100	3	REG	4.7.8	0		
9	7	Interview Exam Room	EXAM ROOM	15	100	1500	1	KPU	6.8	1	0.2.3	
10	8	Workstation Care Team	EXAM ROOM	8	30	240	1	KPU	6.7	1		
11	9	Office Touchdown	PSYCHIATRIC INPATIENT	3	30	90	5	REG	10.11	0		
12	10	Work area Huddle	PSYCHIATRIC INPATIENT	1	80	80	6	REG	9.11.12.13	0		
13	11	Workstation Staff	PSYCHIATRIC INPATIENT	1	100	100	3	REG	8.13.11	0		
14	12	Office Private First Line Leader	PSYCHIATRIC INPATIENT	1	100	100	7	REG	13.11	0		
15	13	Business Centre	PSYCHIATRIC INPATIENT	1	60	60	9	REG	9.10.11	0		

Fig. 6. Requirement supplied to the graph via .csv program document

3.4.2 Generate Building Form

The next step is to construct orthogonal building outline regarding the site layout, the site coverage percentage, and placing departments and programs while developing the form, which will result into rapid form generation options in just a few seconds. The design seed input introduces a random seed influencing the generation of new form proposals. Changing the design seed value will generate a new form option with the same input parameters. The architect will change the design seed value until being satisfied with the generated building form in order to evaluate its environmental behavior in the next step.

3.4.3 Perform Environmental Analysis

As many researchers believe that the solar radiation is one of the most important factors that if considered will lead to energy savings (Ne'eman 1976). Solar radiation is performed using RADIANCE (http://radsite.lbl.gov/radiance/) based on cumulative sky model, which is being used by Ladybug package (https://github.com/ladybug-tools/ladybug-dynamo). For the calculation of the Solar Radiation amount for different sky patches of the Tregenza Skydomes, Ladybug uses the cumulative sky approach (Robinson and Stone 2004). The Ladybug tool has been tested and validated (Roudsari and Pak 2013).

Regarding the importance of the solar radiation, this step is mainly to optimize the orientation of the generated building form. The main input is the energy weather file ".epw" which contains the most important information that is needed for the analysis.

The difference between the total radiation in the summer season and the winter is calculated, then an optimization algorithm using Optimo package will be adopted to determine the optimum orientation for the given form generated from the previous step.

3.4.4 Assign Departments, Programs, and Circulation to the Generated Form

After the architect is being satisfied with the generated form and its optimum orientation, and with the aid of SpaceLayout package, Departments will be assigned into that form based on Key Planning Units and Regular units, considering their importance factors from the Data Stack step, followed by the placement of programs and circulation.

3.4.5 Evaluation of the Generated Space Plans

The SpacePlanFitnessTest node evaluates the generated space layout and conveys the success of design options relative to input goals and constraints based some criteria like Program fitness, defined as the quantity of the required program area successfully assigned by the layout algorithm. The score is computed by averaging the program fitness scores for all departments. Each department's fitness score is computed by dividing the program area achieved in a department by the program area required. The overall program fitness score is returned by the ProgramFitScore output port. When the Designer is satisfied with the total score, he proceed to the last step of that BIM Use workflow.

3.4.6 Exporting Data

After the Designer is satisfied with the total score evaluation of the generated space plan, its form, and orientation, he proceeds to the last step to export the data and geometry of the suggested design to Revit in order to continue with the schematic design phase.

4 Case Study

The BIM Use process was implemented in a hospital case study introduced by Autodesk (see Fig. 7). This process is implemented to enable the design team to generate and explore a large number of design solutions. For this purpose, hundreds of building forms are generated and the selection of five layouts has been done based on satisfying the owner needs. In the next step, for each spatial layout of the selected five options, the optimization of the environmental analysis will be applied considering the solar radiation. An initial population of random (Computer Integrated Construction Research Program 2010) orientation will be generated, then the difference between total solar radiation of the winter and summer will be calculated, the objective function is to maximize that total solar radiation. Finally, the optimum orientation of the selected generated forms will be determined. Assigning the Departments, Programs, and circulation to the generated forms is the next step, then evaluating all of the previous options based on their environmental impact and total score of the design which is a metric that represents the achievement of the generated design to the requirements will be made in order to select the final program design. Exporting the program model to Revit allows the (.ifc) file to be generated which is the standard for BIM exchange.

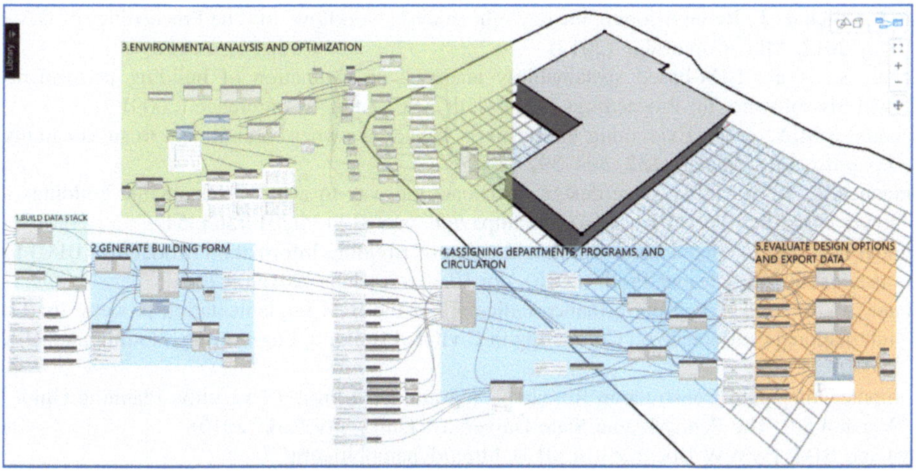

Fig. 7. Case study workflow

5 Conclusion and Recommendations

This paper recognized green BIM as an evolving trend in the AEC industry and identified the need for a standard business process to address BIM execution in green, especially the environmental consideration at the planning stage. This research proposed to introduce a unique BIM execution process model to develop green BIM practices and thus lead to energy savings in buildings. The new process model "Environmentally-based Programming BIM Use" considered the environmental aspects. In addition, it assists designers with a new type of design process - computational design - in which they can simply input their design preferences, generate a broad variety of environmentally-based spatial design, and find the best design solution. The case study was developed on the top of widely used BIM tools "Revit", to make more designers participate in the process.

However, this research has some limitations, it considered only one environmental aspect, Solar Radiation, and its impact on orientation and form. It is recommended for further future work studies to add the parameter of wind flow in conceptual design consideration using the Butterfly package for Dynamo, a validated and open-source engine for Computational Fluid Dynamics (CFD) (project URL at Mostapha et al. 2017).

References

Ali, M., Nassar, K.: Integrating performance and parametric design tools for urban daylight enhancement. In: Proceedings of BS2013, Presented at the 13th Conference of International Building Performance Simulation Association, Chambéry, France (2013)

An, J., Zofchak, J.: Re-envisioning the daylight analysis workflow. In: The Proceedings of DIVA Day 2012. MIT, Cambridge (2012)

Azhar, S., et al.: BIM-based sustainability analysis: an evaluation of building performance analysis software. In: Proceedings of the 45th ASC Annual Conference (2009)

Beheiry, S.M.A., et al.: Examining the business impact of owner commitment to sustainability. J. Constr. Eng. Manag. **132**, 384–392 (2006)

Bragança, L., et al.: Early stage design decisions: the way to achieve sustainable buildings at lower costs. Sci. World J. (2014). doi:https://doi.org/10.1155/2014/365364

Chong, H., et al.: A mixed review of the adoption of Building Information Modelling (BIM) for sustainability. J. Cleaner Prod. (2016). doi:10.1016/j.jclepro.2016.09.222

Coates, P., et al.: The key performance indicators of the BIM implementation process. In: The International Conference on Computing in Civil and Building Engineering, Nottingham, U.K. (2010)

Computer Integrated Construction Research Program: BIM Project Execution Planning Guide - Version 2.0. The Pennsylvania State University, University Park (2010)

Dynamo BIM: [WWW Document] (2015). http://dynamobim.org/

Horman, M.J., et al.: Lean and green: integrating sustainability and lean construction. In: CIB World Building Congress, International Council for Research and Innovation in Building and Construction (CIB), Rotterdam, Netherlands (2004)

Jakubiec, J.A., Reinhart, C.F.: DIVA 2.0: integrating daylight and thermal simulations using Rhinoceros 3D, Daysim and Energyplus. In: 12th Conference of International Building Performance Simulation Association (2011)

Lagios, K., et al.: Animated building performance simulation (ABPS) - linking Rhinoceros/ Grasshopper with Radiance/Daysim. In: Proceedings of SimBuild (2010)

Laiserin, J.: To BIM finity and beyond! (AEC insight column): building information modeling for today and tomorrow (2007). http://www.cadalyst.com/aec/to-bimfinity-and-beyond-aec-insight-column-3686

McNeel, R.: Rhinoceros. NURBS modeling for Windows (2015). http://www.rhino3d.com/jewelry.htm

Mihindu, S., et al.: Digital construction through BIM systems will drive the reengineering of construction business practices. In: IEEE visualization, New York (2008). doi:10.1109/VIS.2008.22

Mostapha, R., et al.: Butterfly [WWW Document]. GitHub (2017). https://github.com/ladybug-tools/butterfly

Ne'eman, E.: Availability of sunshine. Build. Environ. **11**, 103 (1976)

Nguyen, A., et al.: A review on simulation-based optimization methods applied to building performance analysis. Appl. Energy (2014). doi:10.1016/j.apenergy.2013.08.061

Niemasz, J., et al.: Solar zoning and energy in detached residential dwellings. In: Proceedings of the 2011 Symposium on Simulation for Architecture and Urban Design, San Diego (2011)

Paoletti, G., et al.: Designing low energy buildings: application of a parametric tool and case studies. In: The 27th International Conference on Passive and Low Energy Architecture, Louvain La Neuve, Belgium (2011)

Rahmani, A., et al.: Optimo [WWW Document]. GitHub (2015). https://github.com/BPOpt/Optimo

Robinson, D., Stone, A.: Irradiation modeling made simple: the cumulative sky approach and its applications. In: PLEA2004 - The 21st Conference on Passive and Low Energy Architecture, Eindhoven, The Netherlands (2004)

Roudsari, M.S., Pak, M.: Ladybug: a parametric environmental plugin for Grasshopper to help designers create an environmentally-conscious design. In: Proceedings of BS2013: 13th Conference of International Building Performance Association, Chambery, France (2013)

Schlueter, A., Thessling, F.: Building information model based energy/exergy performance assessment in early design stages. Autom. Constr. (2009). doi:10.1016/j.autcon.2008.07.003

Shi, X., Yang, W.: Performance-driven architectural design and optimization technique from a perspective of architects. Autom. Constr. (2013). doi:10.1016/j.autcon.2013.01.015

Stevanović, S.: Optimization of passive solar design strategies: a review. Renew. Sustain. Energy Rev. (2013). doi:10.1016/j.rser.2013.04.028

Subhajit, D., et al.: Integrated spatial-structural optimization in the conceptual design stage of project. In: eCAADe Education and Research in Computer Aided Architectural Design in Europe (2016)

Sustainable Buildings and Climate Initiative: Common Carbon Metric for Measuring Energy Use & Reporting Greenhouse Gas Emissions from Building Operations. United Nations Environment Programme (UNEP) (2009)

US Dept. of Energy: Building Energy Data Book (2012)

Vermeulen, D.: Dynam(o)ite your Design for Engineers – part 5. Revit beyond BIM (2015)

Wang, W., et al.: Applying multi-objective genetic algorithms in green building design optimization. Build. Environ. (2005). doi:10.1016/j.buildenv.2004.11.017

Weytjens, L., et al.: User preferences for a simple energy design. In: PLEA2012 - 28th Conference, Opportunities, Limits & Needs Towards an Environmentally Responsible Architecture, Lima, Perú (2012)

Yan, W., Rahmani, A.: BPOpt: a framework for BIMbased performance optimization. Energy Build. (2015). doi:10.1016/j.enbuild.2015.09.011

The Impact of Urban Fabric on Natural Ventilation for the City of Alexandria

Hagar M. Shalaby[1(✉)], Aia Sherif[2], and Hasim Altan[3]

[1] Department of Architecture, Faculty of Engineering, Alexandria University, Alexandria, Egypt
haidyyy@gmail.com

[2] Sustainable Design of the Built Environment, Faculty of Engineering & IT, British University in Dubai, Dubai, UAE
aia.m.sherif@gmail.com

[3] Department of Architectural Engineering, College of Engineering, University of Sharjah, Sharjah, UAE
hasimaltan@gmail.com

Abstract. Alexandria is one of the most famous cities on the Mediterranean Sea. Many ages have passed over the ancient city and formed huge changes in the urban fabric, until the time came up with the first scheme of Alexandria in the modern age by the British Sir MacLean in 1921. By adopting the principles of the network planning, with perpendicular streets on the coast and parallel to the winds' directions; MacLean was able to create the ideal thermal environment of the coastal city. With water surface (Mahmoudya canal) mediated the fabric and by maintaining the Lake of Hadara, the city achieved the cross natural ventilation effect of the urban fabric level providing the thermal comfort in the deepest point to the south, away from the coast. Unfortunately, many changes have taken place on MacLean scheme during the past ninety years. The lake had been totally replaced with a classy residential area (Smouha) along with huge informal housing expansion to the south-west and east of the city. While Smouha followed a completely contrary pattern, the informal housing expansion came out as the same as the old layout attributes. Smouha became a classy residential hot zone surrounded by informal housing areas enjoying better natural conditions. This paper is an investigative comparison study depending mainly on a fieldwork. The collected air quality numerical measurements of CO, CO_2, levels of humidity and particles matter that came out within the global standards, were used along with the winds' characteristics to assess the real effect of the natural ventilation of the urban fabric. The study has proven that the network planning invented by the early builders of Alexandria has still the preferred layout accommodating the incoming air flow into the city in which the natural ventilation is playing an important role of providing better living conditions and reducing the air pollutants' proliferation.

Keywords: Urban design · Natural ventilation · Urban planning · Air quality · MacLean master plan · Alexandria city

J. Calautit et al. (eds.), *Towards Sustainable Cities in Asia and the Middle East*, Sustainable Civil Infrastructures, DOI 10.1007/978-3-319-61645-2_11

1 Introduction

The Immigration waves from the countryside to the big cities are the main cause of the rapid growth of the size of cities. Of course, this rapid growth was accompanied with the growth of the city's problems that worsened further in the Third World due to the inadequate planning and the very slow response.

The lack of previously prepared plans to confront the urban growth caused huge and unexpected changes in the city shape through history. These changes are very obvious in the Egyptian cities and most represented in the growth of the informal housing and the upper class communities in the desert as well. Alexandria, the most important Egyptian city on the Mediterranean Sea is now completely different from the first adopted master plan by the City Plan Council in 1918, especially regarding the areas now classified as slums or informal housing that cover approximately 3.25% of the total area of Alexandria and mainly located around the downtown areas and characterized by high population density and poor infrastructure.

This dramatic population growth has produced similar growth in services and transportation, and thus doubled fuel consumption as well as the produced pollutants to the atmosphere, without mentioning the growth of factories and their waste. The huge quantity of pollutants is negatively affected the living conditions, resulting in the raising of another form of housing dedicated to the upper class who have the financial ability to provide a better and cleaner life for their families.

In such disturbing situation, the scientific bodies became very concerned with the studies and investigations of the quality of life in the city. Within two years, the university of Alexandria and under the post graduate program of the Department of Architecture, had carried a detailed analytical study of the entrance arteries of Alexandria, where traffic density is the highest as the city is very attractive to residents and visitors from all over the country due its location on the sea and the mild climate.

Fig. 1. The location of the study area in the middle of Alexandria, Egypt (Shalaby 2016)

A big field study and site survey was conducted in three axes: housing and urban - urban agglomerations – informal housing along with studying the natural features of the city and collecting the measurements of the air quality parameters on the main axes of the city (Fig. 1).

2 Historical Background

A hundred thousand city only in 100 years. This is Alexandria where the residents did not exceed the 8000 people in 1800 and became about 400,000 inhabitants in 1900. The population grew even with larger acceleration up to a million and a quarter inhabitant over 52 years only. Now the city is housing close to ten million in 2016.

Initially, Alexandria was planned by Dinocrates who adopted the chess layout planning form focusing on the city's functions. Alexandria gained its highest power in 320 BC as the capital of the empire during the Ptolemaic Greeks. The city continued its flourishing until it became the most famous and the largest city in the world at the time. The orthogonal planning system with streets network vertical on the Mediterranean Sea has continued to be the dominant style, even after the new birth of the city, the new extensions to the east, west and south has adopted the same model.

The reign of Muhammad Ali is considered to be the major shifting point of the urban development of the city. Alexandria has become -since he became the governor and over the following 150 years- the most important port in the Mediterranean and an important centre of foreign trade as the population has increased to 60,000 inhabitants.

Muhammad Ali adopted the most prominent and the first master plan of Alexandria as in 1918, Alexandria master plan project was approved and assigned to the municipal engineer at the time (Sir MacLean) to prepare a general plan for the city, which is known as the scheme of MacLean 1921. Alexandria widened in the 20th century, the urban fabric has greatly expanded to the east, west and to the south directions as the rail way is connecting the whole city. Now the coastline of Alexandria is exceeding 70 km. Due to the rapid growth in the city five planning schemes were adopted:

1. The scheme of MacLean 1921
2. The Master Plan 1985
3. The Comprehensive Master Plan 2005
4. Alexandria Strategic Master Plan 2017
5. Alexandria Development Strategy 2050

3 Methodology

This paper is a comparative study applying investigative approach. The goal is to shape the relationship between the city urban fabric and the air quality through the assessment of the natural ventilation effect on the urban fabric in Alexandria. The study will focus on the main factors affecting the air quality: air pollution by CO and CO_2, particulate matter, heat temperature, relative humidity, noise, wind direction, air flow and air velocity. Defining and evaluating the effect of the NV in Alexandria will be through three phases:

a. Studying the noted differences in the city shape between the first approved master plan 1921 and the current situation.
b. Field visits in order to collect the numerical measurements of the factors affect the air quality in the city.
c. Comparing the air movement between the two situations: MacLean master plan and the current situation.
d. Defining the results.

By comparing the numerical results of the air quality parameters among various areas with different characteristics within the study area, the relationship between the natural ventilation and the urban fabric will be determined.

The study area is extending close to 3 km along the Mediterranean coast, and to the south about 4 km deep. As the region is located in the middle of Alexandria, it contains various forms of the urban fabric -see Fig. 2- that buzzling out the major changes in city shape through history. The following map illustrates the shape of the urban fabric of the study area and the main transportation axes:

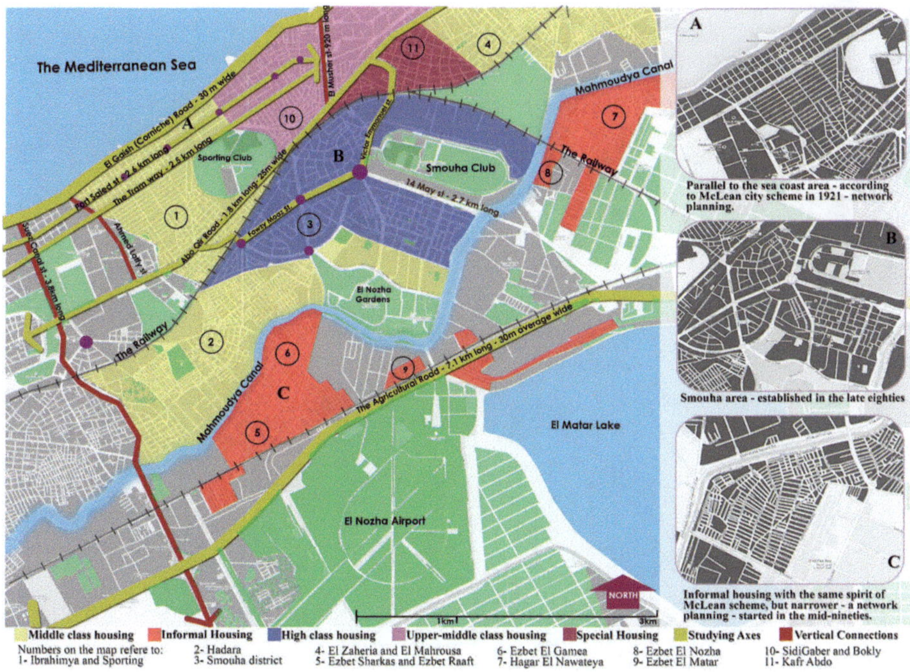

Fig. 2. Study area characteristics; different shapes of the urban fabric within a small region (Shalaby 2016)

4 The Importance of Natural Ventilation Within the Urban Fabric

The modern concept of quality of life is a combination of factors: environment, standard of living, mental and physical health, social position, education, etc. Quality water, quality air and quality food that is what really matters and these basic needs are the foundation on which people build their lives and planners should build their cities (Oleshevych 2015). The growth of the city is paralleled with the growth of transportation. Yet, the risk of increasing the carbon emission becomes larger and preserving the air quality in the city becomes more difficult. Therefore, monitoring the environment is important; environmental indicators provide either aggregate measurements of specific factors (such as total emissions of pollutants over time periods) or quantitative measures of their direct impact on the natural environment itself. The quality of the air we are breathing every day is almost the most important factor of quality of life in the big cities. Therefore, economical techniques and efficient strategies related to natural ventilation should be highly considered in the urban planning (ESE 2013).

Fig. 3. The simple concept of the urban natural ventilation as the situation in the study area of Alexandria (Shalaby 2016)

Natural ventilation is harnessing naturally available forces to supply and remove air in an enclosed space. Whether it is a wind driven ventilation, pressure-driven flows or stack ventilation, the same general principle can be applied to the urban fabric. The urban natural ventilation relies upon the force of the prevailing wind to pull and push air through the enclosed space as well as through breaches in the urban envelope, thus city planning plays the prominent role of the air movement in the spaces (Wikipedia 2016).

Due to the lower wind speed, the higher temperatures, noise and pollution, it difficult to apply the natural ventilation within a wide urban fabric as there is the risk of creating congestion spots in some areas. Because of the specific urban characteristics, there is a serious increase of the ambient temperature and a serious decrease of wind speed in urban canyons. Wind speed within the urban canopy is usually reduced in comparison with rural winds at the same height. The wind speed (u) at any height (z) is lower in the urban areas, and much lower within the obstructed area (Niachou et al. 2007). It is important to provide the appropriate interactions between the city plan and the surrounding regions. A city like Alexandria that enjoys open rural surroundings has better chances to maintain the air movement within the urban fabric. Large variations in the characteristics and the shape of the fabric create more favourable conditions to achieve the natural ventilation in the most closed areas of the city. The map (see Fig. 3) illustrates the north coast zone where wind is produced, the residential zone is receiving the winds and the open back rural region works as a withdrawn machine preserves the continuity of the air movement inside the city.

5 Factors Affecting Natural Ventilation in Alexandria City

As mentioned before, the application of natural ventilation is difficult in the urban environment, especially in street canyons due to reduced wind velocity, urban heat island, noise and pollution. Seven factors were measured among a large range of variation and various types of urban configuration of Alexandria in order to assess the viability of natural ventilation in urban environment (Ghiaus et al. 2005). No doubt that air properties of Alexandria has a significant impact on the natural ventilation. Air flow, air velocity and the wind direction are the main factors determining the success of the natural ventilation.

5.1 Wind Direction

The wind distribution in the urban areas is determined by the prevailing airflow direction with respect to the street axis. Knowledge of the airspeed inside urban streets is of high importance for passive cooling applications (Niachou et al. 2007). The City extends as a long strip along the Mediterranean Sea, the prevailing wind is coming from the north and the north-west (see Fig. 4). These cold damp winds are characterized by Low temperature and contribute to soften the atmosphere of the city. The urban fabric of the coastal city with narrow perpendicular streets to the sea and wide axes parallel to the coast allows winds to penetrate the structural depth (average depth in the study area 3.2 km) of the residential block.

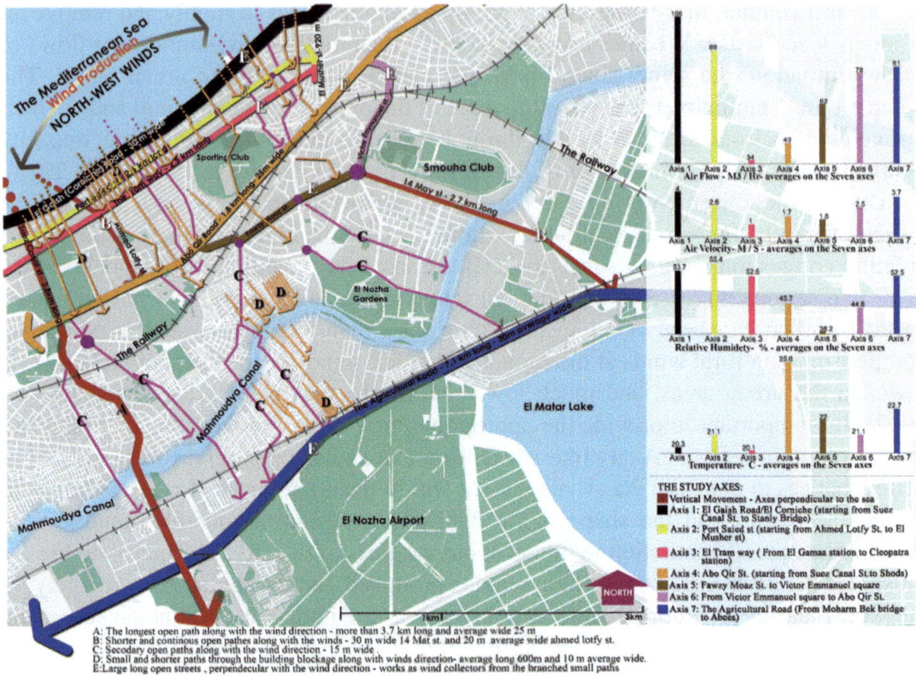

Fig. 4. Wind characteristics - Air flow, Air velocity and wind directions- in the study area (Shalaby 2016)

5.2 Air Velocity and Air Flow

The external fresh air should be delivered to each part of the city in an efficient manner and the airborne pollutants generated in each part of the space should also be removed in an efficient manner (WHO 2009). As the city extends parallel to the sea, it is receiving a large volume of Mediterranean air mass. According to the field measurements the average wind speed in the study area is 2.4 m/s. There is a correlation exists between the undisturbed wind with values larger than 2 to 4 m/s and the wind in the street canyon; when a 2 m/s or stronger wind blows perpendicular to the street canyon, a vortex develops in the canyon. The same winds when parallel to the street become stronger and will be preserved as long as an open path is available (Ghiaus et al. 2005). The coastline is characterized by high speed wind where the highest readings of the air flow were recorded 106 m^3/hr. and air velocity was 4 m/s. Air distribution or airflow pattern changes whenever it permeated the residential block and wind speed become slower, the readings are gradually decreasing to the south until it reaches its lowest level 1.6 m/s in Fawzy Moaz street due to the buildings' blockage (see Fig. 4). It should be mentioned that the lowest recorded winds' readings were 1 m/s and air flow 34 m^3/hr. in the Tramway where the area is below the street level. Winds' speed becomes stronger in the perpendicularly branched streets from the Cornice Road - especially the narrow ones-, forming an effect that is similar to the stack ventilation concept.

5.3 Natural Topography

The urban wind field is complicated. Small differences in topography may cause irregular airflows. As air flows from rural to urban environment, or from the sea to the coastal urban environment depend on the wind direction (Niachou et al. 2007). The topography of the city has significantly contributed to the aeration of the urban fabric; Alexandria had been built on the coastal plain receiving winds directly from the Mediterranean Sea. The effect of these winds is carrying on to the deep south as there are some irregular hills in the southern parts with a height ranges between zero and 40 m above sea level and descend north towards the Mediterranean Sea.

Despite the existence of areas that lies below the sea level in Abu Qir to the east and the fish farms on the southern border of the coastal lagoon, the presence of Lake Mariut works to ease the atmosphere with low-temperature. Moreover, Part of the coastal areas is located between zero and two meters above sea level. The areas range between 3 and 4 m above the sea level are located within the southern part of the delta plain and about 35 km from the coastline. Thus, the high areas enjoy adequate ventilation and a good level of air quality.

5.4 Built-Up Environment - Changes in the Urban Fabric 1921–2016

City scheme and locations of the buildings and spaces are playing a major role in the movement of wind inside the fabric as the structural mass may form a barrier in front of the wind movement. The ventilation strategies are being applied to the buildings can be used in city. In order to provide air quality in the urban areas, the best strategy is to move the static air trapped in certain regions by creating an open access to the sea breeze, this strategy depends on the good quality of the air with minimal pollutants as possible, it is important not to transit the pollutants with the air to the closed areas. Locations of open spaces, green areas, buildings, wide and narrow streets are affecting the wind direction, the ventilation process and then the air quality especially in the far south of Alexandria. In the addition of being the first scheme had been developed for the city, MacLean master plan of Alexandria was featured by the environmental aspects represented in the street network, the green areas and the preservation of the water surfaces. Generally, such aspects were not taken seriously in the following schemes and then the new areas couldn't enjoy the same advantages like before.

The following diagram illustrates the main changes in the MacLean master plan and the current situation according to the following schemes until 2005 (Fig. 5):

- Replacing the Lake of Hadara with the neighbourhood Smouha in 1924 after only 4 years of the adoption of MacLean scheme in 1921 in which the lake was retained as an open area, and before adopting any other master plan of the city. This simple change had the city lose a water body used to cool the air in the southern depth of the city. As Smouha district does not have the same aspects of the street grid in the north, the recorded readings of air flow and air velocity came slower and the pollution readings came higher.

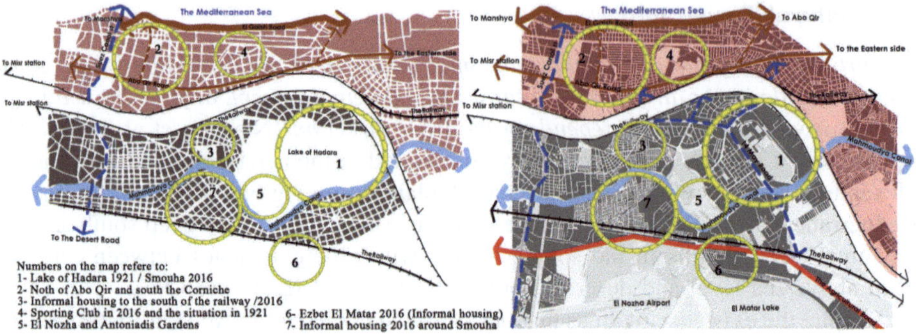

Fig. 5. Main differences between MacLean scheme of 1921-on the left- and the current situation in 2016 – on the right- (Shalaby 2016)

- The north of the city is characterized as the most conservative area in the meaning of preserving the historical aspects and the ancient city's fabric. With the same street network planned according to the wind direction, providing the appropriate light and ventilation.
- Starting from the Nineties, the urban fabric took a different shape; informal housing has begun to appear on the outskirts of the urban areas to the south of the railway. With the sitting up of Smouha, the urban aesthetic of the city has been maintained. However, on the northern east, south and the southern west edges of the new Smouha, slums and unplanned areas have largely appeared.
- The Sporting Club established in 1928 is now occupying very large area; about 97 acres of parks and open spaces. Smouha Club was established in 1949 and so, open spaces increased by 168 acres are the club's area. Until today, Alexandria still maintaining Antoniadis gardens. These green open spaces distributed within the building mass provided the required positive impact on the city ventilation. The polluted air produced in Cornice Road and Abu Qir Street is purified while passing the green mass.
- It existence seemed a fantasy according to MacLean scheme; Ezbet El-Matar is the most known informal housing area where the living conditions are very bad. Although it was difficult to obtain the measurements in the informal housing areas because of the insecurity and people anxiety of the government's actions, according to the site visits it turns out that the area is enjoying a suitable and comfort atmosphere. Ezzab[1] around Smouha were built up with same aspects of MacLean scheme with pedestrianized streets only and even have committed the Egyptian code of buildings' heights; somehow, the low financial conditions of the residents have contributed into creating a good living environment that provides a good air quality. The new Hadara is unplanned area represents the major change between MacLean scheme and the current situation; the differences between the planned wide roads and the random narrow streets are very clear.

[1] Ezzab- the plural of Ezba-: The local name of the informal housing around Smouha district.

The failure to face the rapid growth with strong and clear plans by the State, has led to huge differences in the city form in the modern history. The main goal has changed to house large numbers of the population without serious consideration of the quality of life. After 1921, the city shape changed from a city aim to increase the green and watershed areas to provide a suitable living atmosphere, into a city packed with informal areas and slums, which is negatively reflected on the peace of the city residents.

6 Natural Ventilation Effecting Assessment in Alexandria "Numerical Indicators"

In order to evaluate the air purity degree and the real effect of NV, studying area was divided into seven axes. Seven groups with total of 14 members[2] had collected the measurements of each parameter affecting the air quality. Measurement devices were provided by the architectural department[3]. The results have been used to assess the NV in order to clarify the relationship between the urban fabric and the air quality. Measurements were taken in different locations, they vary from points along the main streets, open fields, highway, and narrow alleys and even in areas not exposed to air flow. The following map illustrates the locations of each reading point (Fig. 6):

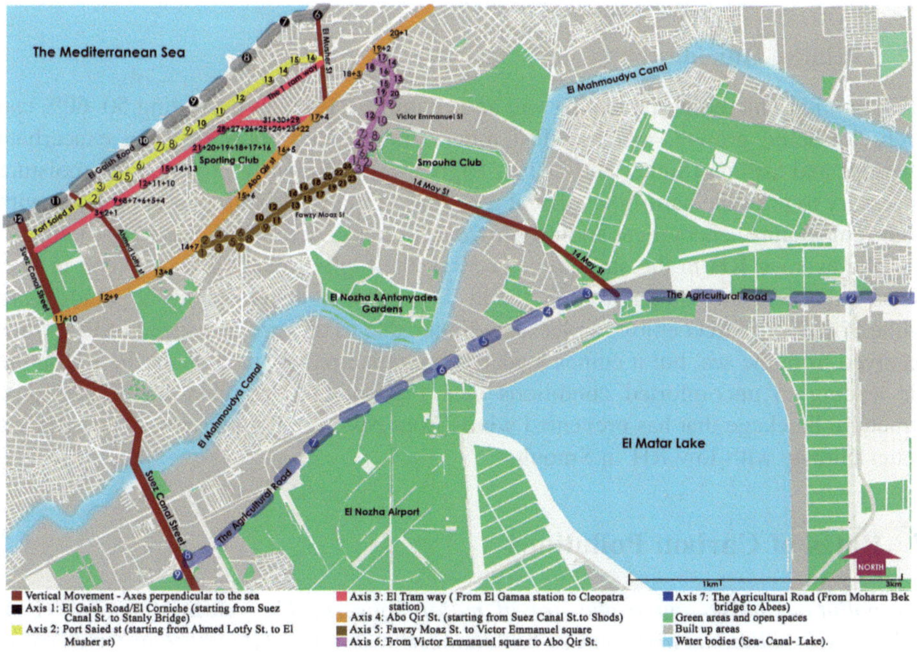

Fig. 6. Locations of the reading points (Shalaby 2016)

---

[2] Post graduate students - Autumn 2014-2015.

[3] Faculty of engineering- University of Alexandria.

6.1 Heat Temperature

The temperature distribution in the urban canopy layer is greatly affected by the radiation balance. Solar radiation incident on urban surfaces is absorbed and then transformed into sensible heat. Most of the solar radiation impinges on roofs and on the vertical walls of buildings; only a relatively small part reaches ground level (Ghiaus et al. 2005). Because the radiant heat loss are slower in urban areas the net balance is more positive than in the surrounding rural areas and thus higher temperature are presented (Niachou et al. 2007).

Warm temperatures may lead to increased frequency and severity of heat-related illnesses and reduced water and air quality along with increasing the risk of cancer, food-borne and waterborne illnesses, cardiovascular and respiratory diseases, diseases from insect bites and spread of diseases from animals to humans, mental health and stress-related illnesses (NCSU 2013).

The highest recorded temperature was 25.6 °C in Abo Qir unlike the Cornice Road where the reading was 20.3 °C as the wind is working to improve the atmosphere and reduce the temperature. Please note that the lowest reading was 20.1 °C in the Tramway although the area has the lowest air velocity and the minimum reading of the air flow. Because hot air is rising to the top and therefore does not actually reach the Tramway below street level (see Fig. 4).

6.2 Relative Humidity

High humidity increases the feeling wet and clammy. RH levels exceeding 50–60% can promote mould growth, increase dust mites, and musty odours that can exacerbate allergies and asthma (Sutherland 2010). High relative humidity provides the moisture necessary to promote harmful chemical reactions in materials and, in combination with high temperature, encourages mould growth and insect activity (Ogden 2016).

The average RH in the study area is 50%, the highest reading was 55.4% in Port Saied street and the lowest was 38.2% in Fawzy Moaz street, which is paradoxical; while Port Saied Street enjoys a pleasant climate due to the size and speed of the winds received from the sea, but it comes loaded with moisture, which is rising the RH in the air leading to uncomforted conditions (see Fig. 4). On the other hand, the same structural blockage that has prevented winds from passing into the street is providing a better climate with low RH in Smouha.

7 Rates of Carbon Pollutions

Air pollution, noise and traffic are of major concern. The relative importance of different air pollutants and sources has changed with time and culture in the different geographical areas. Increased outdoor concentrations seriously affect the indoor concentration of pollutants (Niachou et al. 2007). Outdoor air pollution is commonly considered as another barrier to natural ventilation since filters cannot be used as in mechanical or air-conditioning systems (Ghiaus et al. 2005).

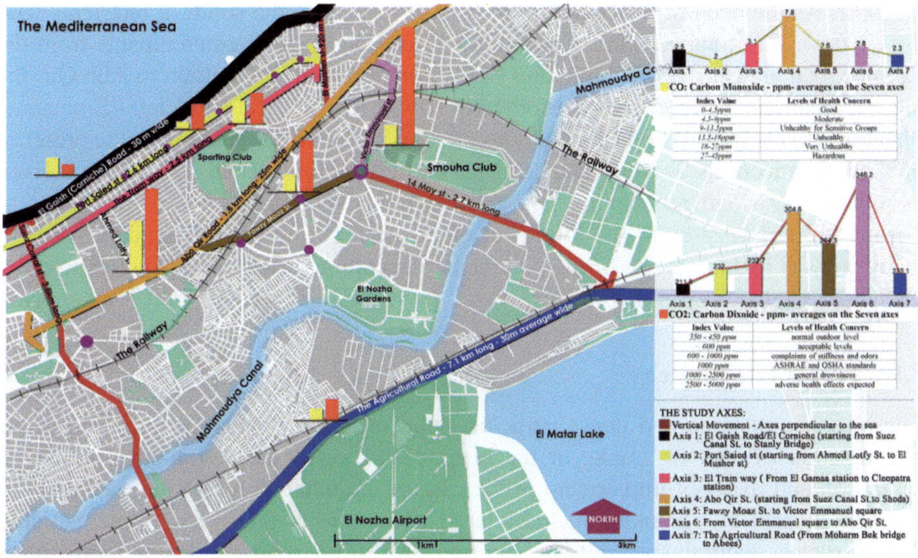

Fig. 7. Averages on the seven studying axes of the most important air pollutants CO and CO_2 (Shalaby 2016)

It is important when analysing pollution to consider that the type and level of pollution differs between closed and open areas, even more, -although initially the economic development increases the carbon levels- when the financial and technological resources become available, the economic growth induces a reduction of the open spaces pollution (Ghiaus et al. 2005).

- **Carbon Monoxide:** CO is produced every time we burn fuel in cars or trucks or even small engines, stoves, lanterns, grills, fireplaces, gas ranges or furnaces. With a long list of risks that varies from breath problems, poisoning to chronic heart disease, anaemia and death, exposing to CO is highly dangerous for everyone. Each year, more than 400 Americans die from unintentional CO poisoning not linked to fires, more than 20,000 visits the emergency room, and more than 4,000 are hospitalized (CDC 2005).
- **Carbon Dioxide:** CO_2 is an abundant greenhouse gas that is believed to be the main cause of global warming and climate change via rising temperatures. CO_2 is a natural bio-product of animal and plant respiration but is also caused by fossil fuel burning, deforestation, cement production and biomass burning. From 650,000 years ago until the industrial revolution the concentration of CO_2 never rose above 280 ppm, in the atmosphere. Since then the level of CO_2 in the atmosphere has grown gradually over time. The CO_2 concentration is increasing at an average global rate of 1.9 ppm per year (NCSU 2013) (Fig. 7).

The map shows that the highest readings of CO and CO_2 were recorded in Abo Qir Street. The differences between readings are huge as it was 2.5 ppm of CO and CO_2 211.9 ppm on the Cornice compared with 7.8 ppm of CO and 304.8 ppm CO_2 in

Abo Qir. As the deep of the building mass is increasing to the south, it interferes with the air flow. The trapped winds in Abu Qir Street loaded with contaminants from the Cornice road are causing the high rates of Carbon in the air. In addition to Abo Qir, the highest readings of Carbon were recorded in Fawzy Moaz Street and Victor Emmanuel Street due to the lack of open connections along with the wind direction as the presence of these connections is important to drain the contaminants outside the city. Even though a very small number of cars are passaging the Tramway and the values of Carbon are not very high, the low level of the path below the surrounding roads in many areas is making it one of the pollutants' deposition spaces as it is not exposed to the common air flows in the city.

7.1 Particulate Matter

The presence of large and dense particles -solid or liquid- in the air may lead to the phenomenon of soot or smoke or the black cloud, respiratory problems, poor concentration and memory impairment in the case of the inhalation of a large quantity of soot and fly ash. This plankton includes large particles as well as microscopic particles, dust particles, natural soil and the fugitive dust from roads, industry, agriculture, demolition, even the flying ash from the combustion of fossil fuels, burning of vegetation and metals smelting and processing.

The highest readings of the particles rates were recorded in Abo Qir, Fawzy Moaz and Victor Emmanuel street due to the continues traffic congestion as theses branched street cannot obtain the large traffic flow coming from the main Cornice road. Fawzy Moaz -1.7 km long- and Victor Emmanuel -1.2 km long- are the only two streets connecting Smouha with Abo Qir and the Cornice; there are not any other open perpendicular streets along with wind direction to ventilate this area. As noted before, as the Tramway is below the street level in some locations, it is exposed to the disposition of polluting particles, which is reflected in the field readings in the following figure (see Fig. 8).

The high rates of polluting particles were recorded on the agricultural road because of the continually waste burning and straw rice fires on the both sides of the road. The large number of trucks and large vehicles -loaded with agricultural products- moving on the road in deteriorating conditions and produce about twice number of pollutants to the air.

7.2 Noise

Noise is one urban form of pollution affecting seriously the quality of life. The most important source of noise pollution is the road traffic, aircrafts and railways (Niachou et al. 2007).

Noise is one of the challenges facing the natural ventilation, air flow pattern transfers the noise from area to another and sometimes create spots of the cumulative noise. The impact of the wind movement on noise transfer is very clear along the Cornice road; the noise rates in the upper floors of residential buildings directly overlooking the Cornice are recording higher readings than on the street level ass the winds are moving up when facing a structural barrier. The highest noise measurements were recorded on the Cornice 80.3 dB through agricultural road 81.5 due to the heavy traffic (see Fig. 8).

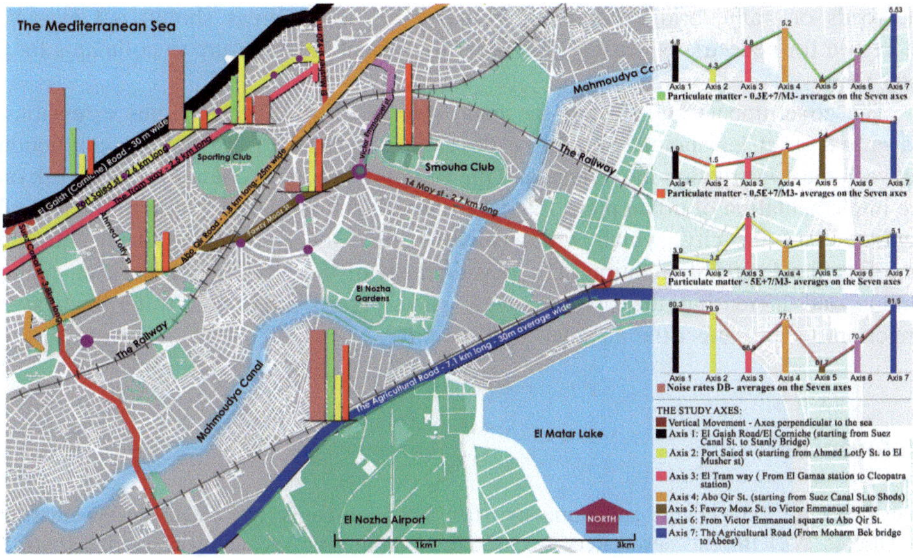

Fig. 8. Averages of the air particles and noise on the seven axes of the study area (Shalaby 2016)

8 Conclusions and Recommendations

It is very important to consider the natural air ventilation in the planning process especially in the coastal areas as the winds properties are dominant. Coastal cities as well as the cities on hills or mountains are enjoying better climate conditions comparing with the inner cities in the flat areas. When urban planning creates spaces that are environmentally closed on itself away from the mutual air flow, it actually creates black spots exposed to the deposition of pollutants. Open paths in the direction of the wind are important to improve the air movement to the depth of the city, but too long paths cause descending in the wind speed and the flow may not reach the depth of the city. The presence of the wide open incidental roads is important for re-producing the scattered air mass and recollecting the air flows. Water bodies or paths within the urban fabric are playing the main role of wind chill. The large green open areas work on reassembling the air mass and cleaning the air from the contaminants.

Adopting the network planning principles has created good ventilated areas in Alexandria; the efficiency of this type of planning has been proved in reducing pollution and creating better thermal conditions in the remote areas away from the Mediterranean Sea, such layout is providing a good air quality with the minimal financial requirements and is very appropriate in planning the middle/low income class.

Coastal cities should be planned to accommodate the incoming air flow into the city, the presence of a long and high solid blockage is act as a barrier against the wind. Heights must be the lowest in the winds reception areas; height increases as the city extend perpendicular to the sea. It is very important not to create pollution collecting spots in the cities; Low lying areas are gathering pollutants if they were located below the roads or traffic bridges in mouth-winds areas. Such areas should not ever be addressed for pedestrians' activities in order to prevent the exposure to pollutants for a long time.

The governments are playing the main role in controlling the cities extensions. Laws and strict legislations should be developed in order to prevent the confliction between the construction movement and the Building law. City council should work on the implementation of urban plans that take into account the preservation of the natural ventilation of the city, which will positively affect the air quality, the health and therefore the financial situation. The development of the city should work on maintaining and increasing the open squares, parks, green areas and water bodies as important factors affect the air quality of the city.

References

Sutherland, A.: Temperature and Humidity, Rain and High humidity = Increased Potential for Disease. AET Insights newsletters, Accredited Environmental Technologies Inc. 1-800-969-6 AET (2010). http://aetinc.biz/newsletters/2010-insights/march-2010. Accessed 15 Sept 2016

Ghiaus, C., Allard, F., Santamouris, M., Georgakis, C., Roulet, C.-A., Germano, M., Tillenkamp, F., Heijmans, N., Nicol, F., Maldonado, E., Almeida, M., Guarracino, G., Roche, L.: Natural ventilation of urban buildings – summary of URBVENT project. In: Proceedings of the 1st International Conference on Passive and Low Energy Cooling for the Built Environment: PALENC 2005, 19–21 May 2005, Santorini, Greece, vol. 1, pp. 29–33 (2005)

Department of Health and Human Services, (CDC) Centers for Disease Control and Prevention: Carbon Monoxide Poisoning (2005). https://www.cdc.gov/co/ar/pdfs/faqs.pdf. Accessed 24 Aug 2016

Eurostat Statistics Explained: Quality of life indicators natural and living environment (2013). http://ec.europa.eu/eurostat/statisticsexplained/index.php/Quality_of_life_indicators_-_natural_and_living_environment. Accessed 25 Aug 2016

Atkinson, J., Chartier, Y., Silvia, C.L.P., Jensen, P., Li, Y., Seto, W.-H.: Natural Ventilation for Infection Control in Health-Care Settings, pp. 7–10. WHO Publication/Guidelines (2009)

Niachou, K., Santamouris, M., Georgakis, C.: Technical Note AIVC 61: Natural and Hybrid Ventilation in the Urban Environment. Annex V Air Infiltration and Ventilation Centre, IEA Energy Conservation in Buildings and Community Systems Programme, pp. 1–10 (2007)

Ogden, S.: The Environment: Temperature, Relative Humidity, Light, and Air Quality: Basic Guidelines for Preservation. NorthEast Document Conservation Center (2016). https://www.nedcc.org/free-resources/preservation-leaflets/2.-the-environment/2.1-temperature,-relative-humidity,-light,-and-air-quality-basic-guidelines-for-preservation. Accessed 28 Aug 2016

Oleshevych, A.: What really determines quality of life? Athens Dialogues E-J. (2015). http://athensdialogues.chs.harvard.edu/cgibin/WebObjects/athensdialogues.woa/wa/dist?dis=92. Accessed 31 Aug 2016

NC State University: Carbon Dioxide (2013). http://climate.ncsu.edu/edu/k12/.carbondioxide. Accessed 30 Aug 2016

Wikipedia: Ventilation (architecture) (2016). https://en.wikipedia.org/wiki/Ventilation_(architecture)#Natural_ventilation. Accessed 25 Sept 2016

Actualization of the Qualitative State of the Water Resources in the Plain of Gharb (Downstream of Sebou - Morocco)

Nezha El Mahmouhi(✉), Mohammed El Wartiti, Rajae Azelmad, Ridouane Sadki, and Sanae Berred

Department of Earth Sciences, University Mohammed V, Rabat, Morocco
mahmouhnezha@yahoo.com, wartiti@hotmail.com, rajazelmad@gmail.com, r.sadki@hotmail.fr, berredsanae@gmail.com

Abstract. The Gharb plain (Northern Morocco) benefits from a privileged geographical situation, very fertile soils, and relevant water resources. Geochemical characterisations of the groundwater and of the water-rock interaction processes are suggested. A shallow aquifer and a deep multi-layered aquifer have been distinguished. The shallow aquifer is hosted in silty clay formations, it supplies an alkaline-chloride type of water with the total dissolved solids (TDS) in the 0.7–1.8 g/l range, and the dissolved contents depend on the dissolution of sea salts, calcite, gypsum, and dolomite; salt domes leaching and cationic exchange or inverse exchange processes may occur locally, increasing the TDS up to 4 g/l. Less salty water (TDS: 0.2–0.9 g/l) comes from the multi-layered aquifer, and shows a calcium bicarbonate composition related mainly to the interaction with carbonate minerals, which prevail in the sediments. All waters are over-saturated with respect to calcite, and under-saturated with respect to gypsum.

Keywords: Gharb plain · Morocco · Geochimical · Ground water

1 Introduction

The Gharb plain is located in the Atlantic coastal zone at the junction of two structural domains of north-western Morocco. The plain extends on a surface of 4000 km^2, and represents about 50% of the total surface of the Gharb - Maamora hydrogeological basin (Fig. 1). During the past decades it has witnessed a population explosion (1,500,000 inhabitants with an average annual growth evaluated at 3%). The Gharb plain benefits from a privileged geographical situation in the Kingdom of Morocco, very fertile soils, and important water resources. The renewable water resource in the area is supposedly 150 million m^3/y, and the reservoir/renewable resource ratio amounts to 130 years. These data show the importance of this resource for a semi-arid country like Morocco in a period of increasing drought. This study aims at a preliminary geochemical characterisation of the groundwater in order to set out the water-rock interaction processes as a premise to distinguish natural phenomena from others linked to the anthropogenic contamination of the water. A campaign to measure the

J. Calautit et al. (eds.), *Towards Sustainable Cities in Asia and the Middle East*, Sustainable Civil Infrastructures, DOI 10.1007/978-3-319-61645-2_12

physico-chemical parameters and to sample the water has been carried out. The data from the field and laboratory investigations have been integrated with data on the lithology of different drillings, gathered from the General Management of Hydraulics (GDH) and from the Management of Water Research and Planning (DRPE) of the Kingdom of Morocco.

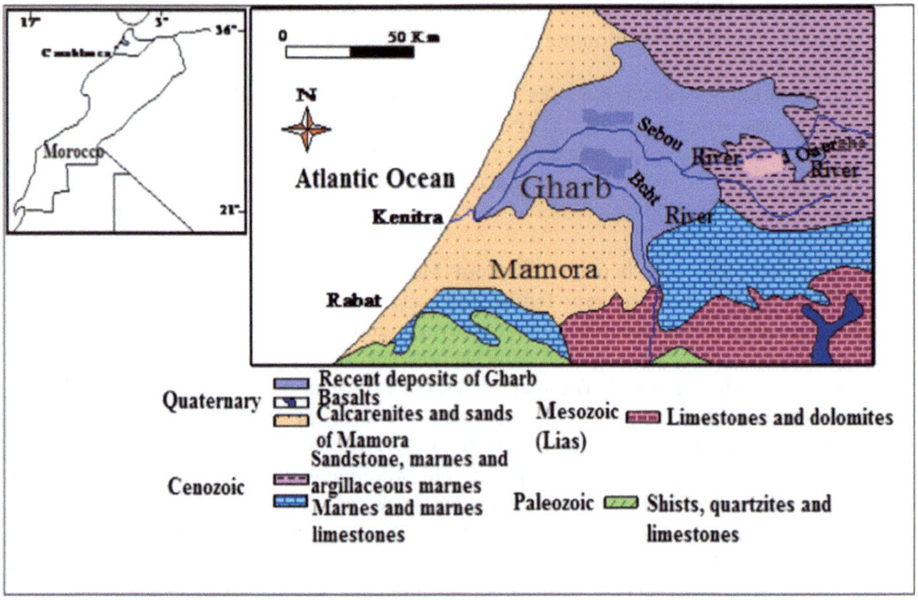

Fig. 1. Situation map of the geology of Gharb plain and its borders

2 Previous Studies

The hydrogeochemical characteristics of surface and shallow waters of the Gharb plain are reported in previous studies on the geochemistry of water carried out on the entire basin of the Sebou river (Caboi et al. 2000; Saddiki et al. 2001; Bahaj et al. 2001) and in a hydrogeological study with Fe and Mn data from the water of Gharb (Kacimi 1992). This paper presents a more detailed study on the geochemistry of shallow and deep water in the Gharb plain.

3 Geology

Two important geomorphological units are distinguished in the Gharb basin: (i) the plain, which represents 90% of the basin, has a triangular form and is crossed by the Oued Sebou and its tributaries (Ouergha, Beht); (ii) the coast formed by a system of sub-parallel dunes which separate the plain from the Atlantic Ocean (Fig. 2). The Gharb is limited to the north by Mesozoic and Tertiary formations of the Rif, folded during the Alpine orogeny, and to the southern-west by the Moroccan Meseta, made up of the Hercynian basement. The subsidence of the geological basin hosting the Gharb aquifers is related to the same tectonic phenomena that give origin to the Rif Chain (Michard 1976). The bottom of the basin, consisting of a tectonic nappe dating back to the Miocene, is made of detrital clay, turbidites, and blue gypsum-marl with interbedded Triassic evaporites. The Middle Miocene is represented by limestone of a thickness of 150–200 m (Cirac 1987). In the upper Pliocene, the sedimentation is characterised by coastal type deposits: calcarenite sand and sandstone in the southern-east, and conglomerates in the north. During the Quaternary, the first transgression is marked by sea deposits essentially made up of calcarenites (Aberkan 1989). This coarse and permeable material makes up the water reservoir of the deep aquifer hosted in the coastal dune complex. Later Quaternary transgressions originated other dunes following a parallel strip in the beach with a sedimentation of essentially marine origin.

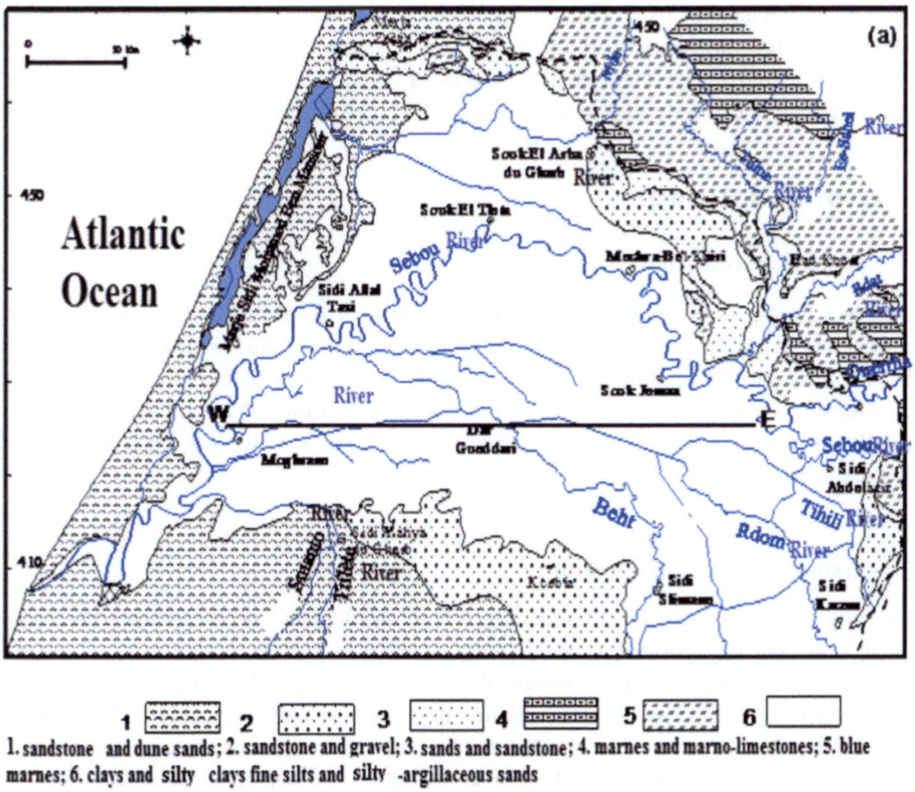

Fig. 2. Schematic map of the lithology outcropping in the Gharb-Maamora hydrogeological basin (in Kacimi 1992, modified).

4 Hydrogeology

Two aquifer units have been distinguished: (i) the shallow aquifer spreads over the entire plain, except at the borders of the basin in which a deep formation outcrops; (ii) the deep multi-layered aquifer.

Shallow water is contained in superficial silty clay with low permeability. The upper part of the reservoir is made of clay with a permeability of between 1.10^{-7} and 5.10^{-7}, or of sandy silts with a permeability of between 1.10^{-7} and 4.10^{-6} m/s. In the coastal zone of the plain, the flow direction is mainly towards the Atlantic Ocean. The system of deep multi-layered aquifers consists of coarse levels of gravels, and sands whose permeability varies between 1.10^{-4} and 6.10^{-3} m/s, while in the west part the sediments of the aquifers show a finer granulometry, essentially made up of sands, silt, and sandy clays whose permeability is estimated between 1.10^{-5} and 5.10^{-5} m/s. The multi-layer character is related to the presence of lenses of permeable coarse sandy material including pebbles, limestones, calcarenites, and sandstones, at times separated by layers of clay. The flow in the deep aquifer is prevalently directed towards the centre of the plain, because of its pan shape, and towards the Atlantic Ocean only on the west side. Except for the borders of the basin where deep formations outcrop, the deep aquifer is covered by an impermeable high layer of clay. Therefore the recharge of the aquifer, which is estimated up to 150 million m^3/y, mainly occurs at the basin borders.

5 Hydrogeochemistry

5.1 Sampling and Analytical Methods

Forty-one wells, used for irrigation and the supply of drinking water, randomly distributed on the plain, were sampled in September 2001 and January 2002. Temperature, pH, redox potential, conductivity, alkalinity, NH_4^+, and NO_2 species were measured at the sampling sites. In the laboratory, anions were determined by ion chromatography on filtered, non-acidified samples. Cations were analysed by ICP-OES on filtered samples acidified to 1% HNO_3.

5.2 Groundwater Classification

Generally, the water of the different aquifers of the Gharb varies between the bicarbonate type and the chloride type for anions, and between the calcium type and the sodium type for cations. Most waters of the deep aquifer, characterised by a bicarbonate majority for anions and a calcium majority for cations, can be classified as calcium bicarbonate water. The shallow water can be classified as the alkaline-chloride type with a variable alkaline-earth ion contribution of between 20 and 30%. The variation in the concentration of the major ions in relation to total salinity (TDS: total dissolved solid) in the Gharb aquifer (Fig. 3) shows that the water in the deep aquifer is less salty (TDS: 0.24–1.2 g/l).

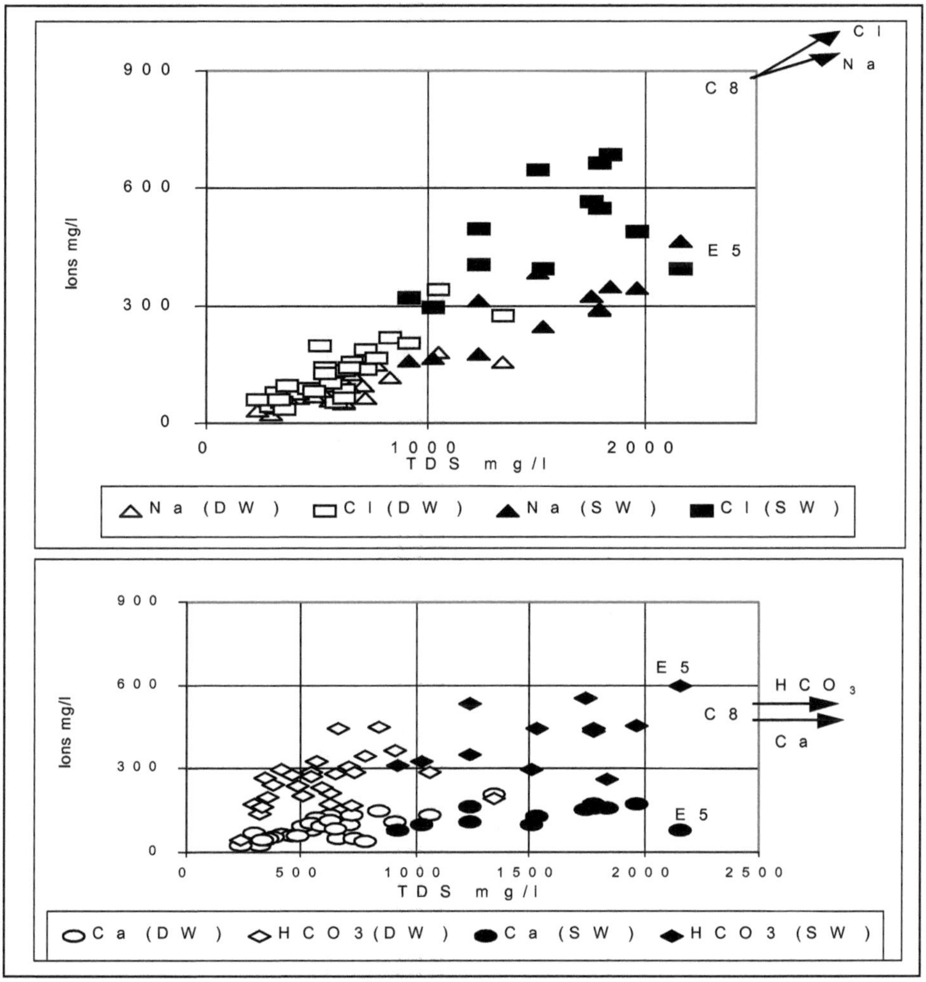

Fig. 3. Concentration of main cation and anion vs TDS in the shallow (SW black symbols) and deep water (DW open symbols) samples of the Gharb aquifers.

Bicarbonate for anions and calcium for cations give the greatest contribution to salinity.

On the contrary, the water of the shallow aquifer is the saltiest (most samples in the range: 0.9–1.8 g/l). Two samples, E5 and C8, with respectively TDS = 2.2 and 4.5 g/l reveal a few peculiarities, which will be discussed in the next section. Sodium and chloride predominate, while sulphate and magnesium are always subordinated to other ions.

5.3 Water Geochemistry

The diagrams in Fig. 4 help distinguish the different mineral-water interaction processes in the soils and the sedimentary aquifers. Most samples, both in deep and

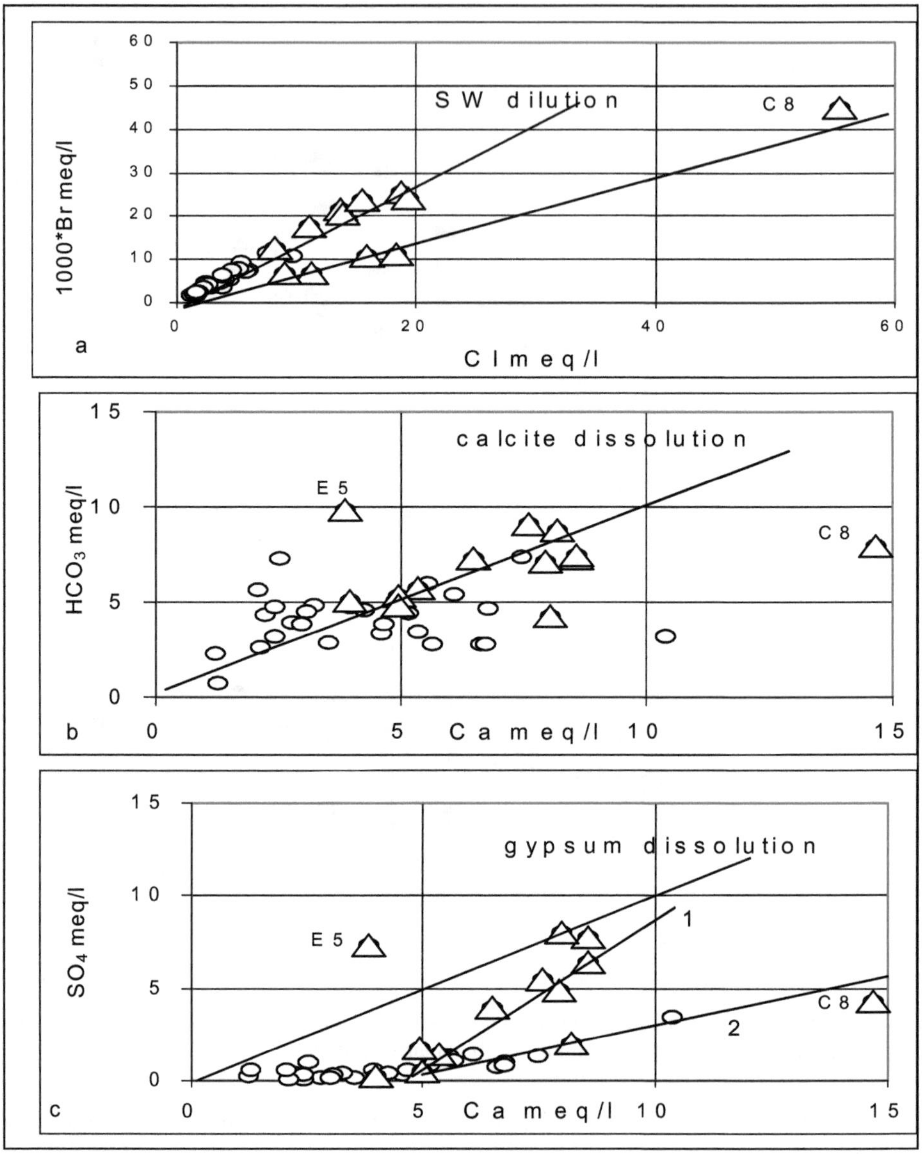

Fig. 4. (a, b and c) Shallow and deep water evolution lines (see text; circles: deep water; triangle: shallow water)

shallow water, lie on a line with a Br/Cl ratio that can be referred to the sea water dilution line (Fig. 4a). Another trend with a lower Br/Cl ratio than that of sea water can be discerned and can be referred to halite dissolution from salts domes (Hounslow 1995) that are widely spread in the area. Calcite dissolution is among the main processes of mineral-water interaction (Fig. 4b); the calcium deficit (about −5 meq/l with

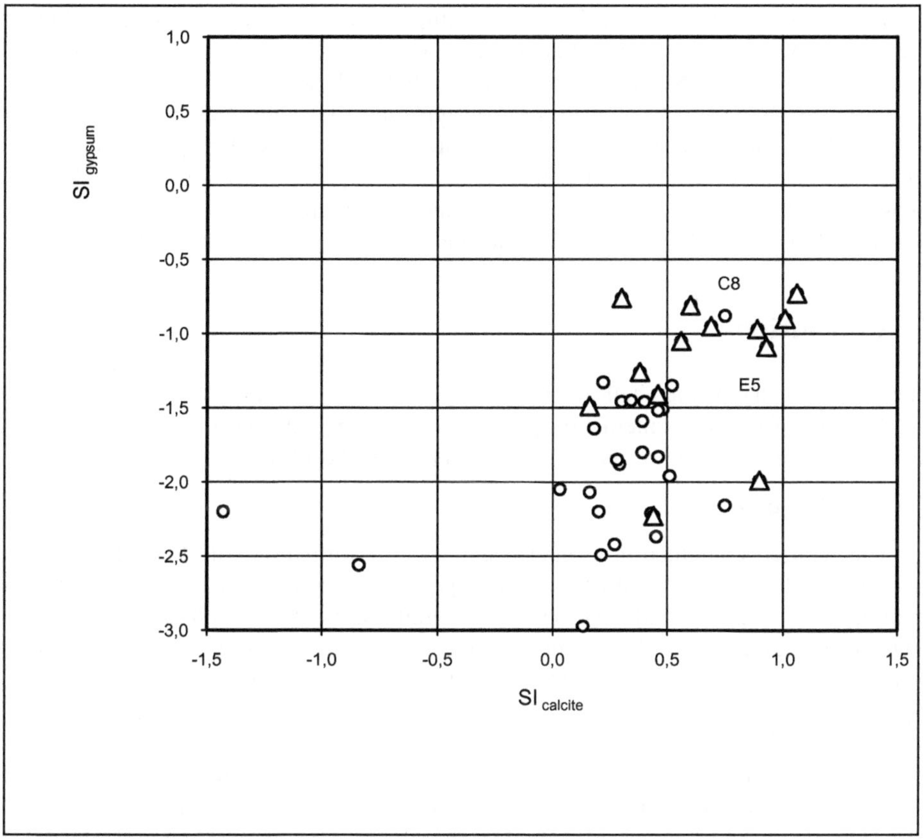

Fig. 5. calcite and gypsum saturation indices in the shallow and deep water of Gharb plain; (symbols as in Fig. 4)

respect to other shallow waters lying on the calcite dissolution line) in the "E5" type water can be attributed to the cationic exchange reaction:

$$5Na_2{-}montm + 5Ca^{2+} -> 5Ca{-}montm + 10Na^{+}$$

Concordant with the +10 meq/l Na excess with respect to most shallow waters. Some other deep waters also underwent a similar process. The different trends observed in the SO_4 vs Ca diagram (Fig. 4c) show that: (i) gypsum dissolution is not the predominant mineral-water process; (ii) the two evolution lines of the shallow water suggest a gypsum plus calcite (and dolomite) dissolution for trend 1 and the overlapping of this process with a reaction of inverse cation exchange for trend 2 as:

$$Ca{-}montm + 2Na^{+} -> Na_2{-}montm + Ca^{2+}.$$

Trend 2 also includes some deep water samples and especially sample C8, which is the saltiest, whose Br/Cl ratio points to an interaction with evaporitic rocks which had caused an increase in Na concentration due to halite dissolution.

In Fig. 5 SI_{gypsum} vs $SI_{calcite}$ are shown (gypsum and calcite saturation indices, SI = log (IAP/K_s)). Most shallow and deep waters are oversaturated with respect to calcite and all the samples are undersaturated with respect to gypsum. The saturation indices of gypsum for the shallow water do not exceed values of about $SI_{gypsum} = -1.0$. This could be referred to a control of the calcium concentration caused by calcite precipitation.

6 Conclusions

The geochemistry of the shallow and deep waters of the Gharb plain depends on the overlapping of different interaction processes of water with the minerals of the sedimentary marine deposits of the Lower Miocene and the Pliocene.

(1) ***Shallow aquifer***

This aquifer, contained in a superficial silty clay formation with low permeability, can be classified of the alkaline-chloride type with a variable alkaline-earth ion contribution of between 20 and 30%. TDS is generally in the 0.9–1.8 g/l range, depending on the dissolution of sea salts, calcite, gypsum and dolomite; at times cationic exchange or inverse exchange processes, which raise the TDS to 4 g/l, can occur.

(2) ***Deep groundwater***

The deep multi-layer aquifer supplies less saline water (TDS: 0.2–1.2 g/l) of the calcium bicarbonate type (most of which) oversaturated in calcite, linked to interaction with limestones, calcarenites, and conglomerates in the eastern part and in the North of the plain. Interaction with the layers of clay minerals interbedded in the permeable levels locally causes a Ca/Na or a Na/Ca exchange.

Acknowledgements. This study was supported by funds from the Sardinian Regional Government (RAS).

References

Aberkan, M.: Studies of the quaternary formations of the margins of the basin of Gharb (Morocco, North-West). Doctorate of state University of Bordeaux, vol. 1, no. 964, 290 p. (1989)

Bahaj, T., El Wartiti, M., Zahraoui, M., Caboi, R., Cidu, R.: Shallow groundwater in the Sebou basin (Northern Morocco). In: Proceedings of the 10th International Symposium on Water-Rock Interaction -WRI-10, Villasimius (Italy), pp. 445–447. Balkema, Rotterdam (2001)

Caboi, R., Cidu, R., Carcangiu, M., El Wartiti, M., Zaharaoui, M., Bahaj, T., Saddiki, N., Bahaj, S.: New hydrogeochemical data in the Oued Sebou basin (Morocco). In: Proceedings of the 5th International Conference on the Geology of the Arab World, pp. 226–227. Cairo University, Egypt, February 2000

Cirac, P.: The south-west basin with the superior Neocene. Evolution of sedimentary dynamics and paleogeography during a phase of filling. Mem. Inst. Geol. Basin of Aquitaine. University of Bordeaux, vol. 1, no. 21, 287 p. (1987)

Hounslow, A.W.: Water Quality Data: Analysis and Interpretation. Lewis Publishers, Boca Raton (1995)

Kacimi, I.: Some aspects on the hydrogeology and the hydrochimy of the water aquifer of Gharb - Morocco. The particular problem of iron and Manganese. Control in earth sciences, 154 p. University of Quebec, Montreal (1992)

Michard, A.: Elements de Moroccan geology. Notes Mem. Serv. Geol. Morocco, Rabat, no. 252, 408 p. (1976)

Saddiki, N., El Wartiti, M., Zahraoui, M., Caboi, R., Cidu, R.: Anthropogenic impact on the quality of surface waters in the Fes-Meknes plateau (Morocco). In: Proceedings of the 10th International Symposium on Water-Rock Interaction -WRI-10, Villasimius (Italy). pp. 1135–1338. Balkema, Rotterdam (2001)

Solution Model for Infrastructure Planning and Design for Urban Slums: Egyptian Case Study

Mark Sedrak(✉), Yasser El Khaldi, Safwan Khedr, Maram El Saudy, Emad Emam, and Ahmed El Gendy

The American University in Cairo, Cairo, Egypt

Abstract. Urban Slums represent a chronic problem to Egypt. According to official statistics, there are 1,221 urban slum inhabited by 17 million people in Egypt. These urban slums suffer from deterioration in infrastructure; this includes lack of access to proper sanitation, inadequate water distribution networks and congested street networks. These issues in infrastructure systems are reflected on the social aspect and the quality of life in these areas with problems such as high crime rates, health problems and spread of epidemics.

This study aims at presenting a solution model for improving the infrastructure networks in urban slums taking into consideration the urban, engineering and network design aspects. Suitable guidelines and bases of design are set in a solution model that can be revisited by studies and projects which encompasses infrastructure development in urban slums. The model considers different factors of social, economic and engineering aspects that may affect efficient and effective solutions.

A case study presented in this study is Al-Sharabiya that is one of the low-income neighborhoods located in North Cairo. This neighborhood suffers severely from social, political and economic issues including uncontrolled population growth rates, low-income levels, high crime rates, drug abuse, high unemployment rates and deteriorated public services such as health, education, and transportation. The Egyptian Planning Authorities focused on this area and initiated an urban development and renewal project to enhance the urban fabric of the neighborhood. In this study the presented model is applied on the proposed urban development maps. The defined scope of this case study is divided into three main components for infrastructure systems design that encompasses transportation network planning and roads design, water distribution network, and sewage system network. The infrastructure systems are designed to improve the living conditions of the inhabitants of the neighborhood through providing adequate transportation, water and wastewater network systems.

Keywords: Infrastructure development · Urban slums · Solution model · Transportation · Water and wastewater networks

1 Introduction

Informal developments have been and continue to be the main mode of urbanization in Egypt. The rate of urbanization has increased from 50 million, and it estimated to grow to 57 million by the year 2025. Failure of public policies to provide adequate housing

J. Calautit et al. (eds.), *Towards Sustainable Cities in Asia and the Middle East*, Sustainable Civil Infrastructures, DOI 10.1007/978-3-319-61645-2_13

for low-income population has led to an increase in the number of people inhabiting informal settlements. In 1998 informal settlements sheltered more than 7 million inhabitants while in 2006 number grew to shelter about 10.5 million from a total of 16.2 million inhabitants living in Cairo. The previous numbers indicate the need for finding a solution for such a chronic problem. The proposed developed model is an attempt to contribute in solving this problem. Al-Sharabiya, located in North Cairo, is chosen as a case study for the proposed model because it suffers severely from economic, social and political issues such as low-Income, high crime rates (Three zones are characterized as being "unsafe" by the authorities), drug abuse, high unemployment rates and deteriorated public services such as health, education and transportation. The model should be able to propose practical and implementable infrastructure network designs in urban slums taking into consideration the urban, engineering and network design aspects. The infrastructure networks discussed in this paper are the transportation network, water distribution network and wastewater collection network. The model's effectiveness and the proposed solutions are thoroughly discussed in the case study.

2 Developing a Solution Model for Infrastructure Planning and Design for Urban Slums

Figure 1 shows the proposed model that offers solutions to the process of infrastructure planning and design in urban slums. The first step is to identify the objectives of the infrastructure development; which comprises the infrastructure networks to be designed as mentioned above. This step is followed by identification of the study scope; this comprises both the characteristics of the study area and the model parameters. The model parameters are classified into 3 main categories, which are the urban planning, engineering and network design parameters. After collecting the data regarding the parameters and the study area, the networks are designed and simulated using appropriate tools/software; the previous process is an iterative one where suggestions are made for improvement till an accepted performance is achieved.

2.1 Identifying the Objectives of the Infrastructure Development

The starting point is to identify the objectives of the infrastructure development; in this model three main infrastructure networks are studied, which are the transportation, water distribution and sewer networks. The transportation network was selected in this model as Cairo suffers from many problems regarding its traffic as traffic congestion, car accidents, and pedestrian accidents. In order to develop an urban slum it has to have an efficient transportation system in order to attract investments in the area and improve the quality of life of its residents. The water network and the wastewater collection system were selected in order to improve the accessibility of clean water and proper sanitation to all people living in urban slums. Such improvement is critical to mitigate the spread of diseases related to lack of clean water and sanitation such as Hepatitis C.

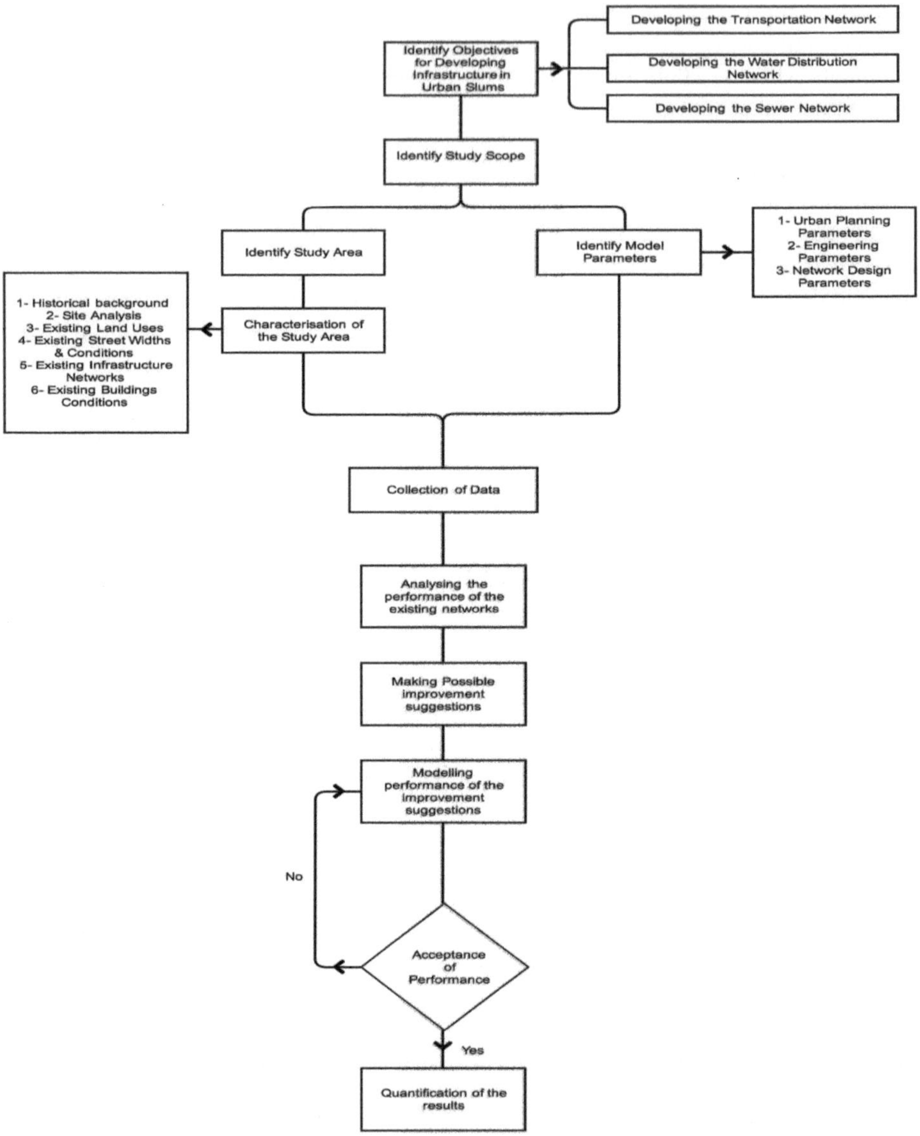

Fig. 1. Solution model for infrastructure development in urban slums

2.2 Identification of the Study Scope

Within the step of identifying the scope of the study both the study area and the study parameters have to be addressed.

2.2.1 Identification of the Study Area

This step involves the characterization of the study area; including the following aspects: Historical Background of the area, cultural, social and economic conditions, demographics and census data, site analysis with the main landmarks and development projects near the area, existing land uses, existing street widths and conditions, existing infrastructure networks and existing building conditions and heights.

2.2.2 Identification of the Main Influencing Parameters

The crucial parameters for the infrastructure planning and design solution model are classified into three main categories: urban planning, engineering and network design parameters. These parameters are the basis upon which optimized solutions in the network design can be made.

(a) The Urban Planning Parameters include: The urban development vision, implementation process of the development project, population density, land use map, right of way (ROW), building heights according to the new (Floor To Area Ratio or FAR) guidelines, street widths, infrastructure networks, trip generation rates, mode choice, vehicle occupancy factors and growth factors
(b) The Engineering Parameters (The inputs to the network design process) include:

- *The Topography of the Site*: Topography of the site is the first engineering parameter, which has to be addressed to design the different proposed networks. Engineering parameters for each of the three proposed networks are discussed in details in the following paragraphs.
- *Transportation Network Engineering Parameters:*
  - For Traffic Planning, the engineering parameters include friction factor and passenger car equivalent factors.
  - For Cross Section Design, the engineering parameters include design speed, capacity of roads, volume to capacity ratio, level of service, adjustment factors and service volumes.
  - For Horizontal Alignments, the engineering parameters include PIEV time, minimum safe radius/maximum safe curvature, lateral clearance for Stopping/Passing Site Distances and length for Super elevation Runoff.
  - For Vertical Alignments and Drainage Network Design, the engineering parameters include Maximum and Minimum Grades and Stopping/Passing Site Distances.
  - For Intersection Design, the engineering parameters include Minimum radii for intersection curves.
  - For Asphalt Pavement Design, the engineering parameters include Load Equivalency Factors, Resilient Modulus (Mr) for asphalt layers, Structural Layer Coefficients (a) and Drainage Coefficient (m).
- *Water Network Engineering Parameters:*
  - These include Average water consumption rates, Peak Hourly Factor and Junction elevations.

- *Sewer Network Engineering Parameters*
  - These include Average Wastewater consumption rates, Minimum Allowable Slopes of sewer lines, Minimum, Maximum and proportional velocity and Minimum depth of sewer lines.

(c) The Network Design Parameters (The output of the design process)

- *Transportation Network Design*
  - For Traffic Planning network design parameters include Equivalent Passenger Car per Hour (EPCPH) on each link.
  - For Cross Section Design network design parameters include Number of Lanes, Lane Width, Median Width, Shoulder Width and Cross Slope.
  - For Horizontal Alignments network design parameters include Point of Curvature, Point of Tangency and Point of Intersection.
  - For Vertical Alignments and Drainage Network; network design parameters include Vertical Point of Curvature, Vertical Point of Tangency, Vertical Point of Intersection and Catch Basin Map.
  - For Intersection Design network design parameters include Radii for intersection curves and Level of Control.
  - For Asphalt Pavement Design network design parameters include Surface Course, Base Course and Sub base Course Thicknesses.
- *Water Distribution Network*
  - These include Pipe Diameters, Pipe Materials, Pressure in Junctions and Flow Velocity.
- *Sewer Network*
  - These include Skeleton layout of the sewer network, Conduits Diameters, Conduits Slopes, Conduits Materials, Crown or invert elevations, Flow Velocity and Flushing System.

2.3 Collection of Data

After the study scope is identified, data is collected regarding the important parameters that were identified in the previous step. These data include the Egyptian Codes of Practice for the design guidelines of each network, commercially available pipe diameters for water and sewer networks, existing infrastructure networks and their conditions such as the main sewer lines and the main water pipes supplying the neighborhood and any other missing information needed for the design process. It is crucial to gather accurate information to guarantee the model's success and the efficiency of its output.

2.4 Modeling the Existing Networks Using Appropriate Software

In case sufficient information regarding the existing infrastructure networks is provided, the performance of each network is simulated using appropriate software. Figure 2 shows the infrastructure design elements with the suggested software for the design of each network.

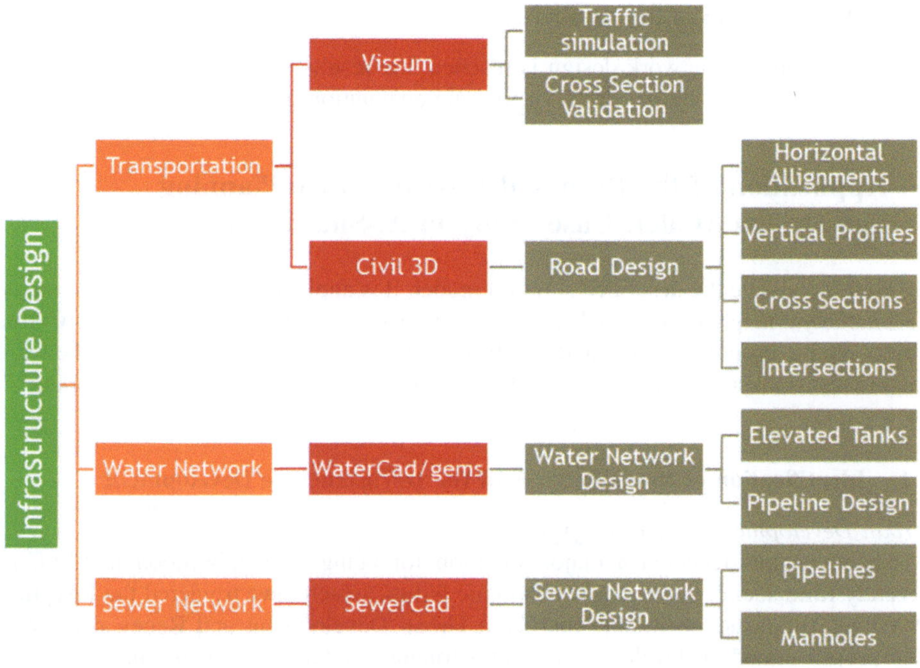

Fig. 2. Infrastructure design elements with the suggested software

2.5 Analyzing the Performance of the Existing Networks

After the existing networks are modeled, the performance of each network is analyzed to identify the areas that need development and improvement. In addition, the analysis of the performance may include field visits and site surveys in order to develop a clearer perspective for the problems that needs development.

2.6 Making Possible Improvement Suggestions

Based on the previous steps, possible improvement suggestions are made to the network design. This step may involve proposing different scenarios to improve each problem identified.

2.7 Modeling the Performance of the Improvement Suggestions

The proposed improvements are modeled and the different scenarios are analyzed and reiterated till an optimized design solution is reached for each problem in each designed network.

2.8 Quantification of the Results

Once an acceptable network design is reached, the results are quantified in tables and diagrams, which will be the basis for the implementation of these designs.

3 Application of the Proposed Infrastructure Planning and Design Model: Case Study in Al-Sharabiya, Egypt

In order to validate the described solution model, it is applied to the selected case study El Sharabiya neighborhood, which is a low-income informal settlement in Cairo. Solutions are proposed, simulated, analyzed till an acceptable design is reached for each of the transportation, water distribution, and wastewater collection networks.

3.1 Identification of the Objectives of the Infrastructure Development

Urban Development Project Purpose

Al-Sharabiya has received a major attention for being a neighborhood that suffers severely from social, political and economic issues; which have triggered the Egyptian Planning Authorities to initiate the Cairo Urban Development and Renewal Project. The project aimed at developing the infrastructure systems and increasing its accessibility to reach all of the slums and areas in this neighborhood in order to enhance the urban fabric of the area. The development will lead to higher land prices; which will be an incentive for people to sell their lands and at the same time investors would be motivated to start up projects in the developed area. Moreover, it will improve the living conditions of the inhabitants of Al-Sharabiya Neighborhood.

Objectives of the Development

The infrastructure development aims to develop three major networks, which are the transportation, water, and sewer networks.

- *The Development of the Transportation Network Aims to:*
  - Reduce the traffic congestion
  - Widen the streets to fit in the water and sewer lines
  - Provide parking lanes
  - Provide safer horizontal curves
  - Reduce the accumulation of water in the street in case of rain
  - Provide a drainage network to remove excess water in the streets
  - Improve the quality of the asphalt pavement
  - Provide safer intersections for cars and pedestrians
- *The Development of the Water Network Aims to:*
  - Improve the Accessibility of All Residents of the Area to Clean Water
  - Fit in a water network that can reach buildings in narrower streets
  - Provide adequate water pressure at each junction that can reach the designed building heights

- *The Development of the Sewer Network Aims to:*
  - Improve the accessibility of all residents of the area to proper sanitation
  - Provide an efficient design without needing to fit in pumps
  - Propose an efficient flushing system

3.2 Identification of the Study Scope

3.2.1 Identification of the Study Area

Neighborhood Historical Background

Al-Sharabiya is a low-income neighborhood located in North Cairo. Historically, it was inhabited by people working as "Water providers" or "Shorraba'ah" in Arabic in the era of Mohamed Ali Pasha who ruled Egypt at that time. Another reason for its name is that the name of the army group, which was responsible for transporting the supplies to the army camps in the Fatimid period, was "Al-Sharabiya".

With regards to the administration responsible for the neighborhood, Al-Sharabiya became independent from Shobra neighborhood in 1980s to become united with El Zawia El Hamra neighborhood and then they were separated again in late 1990s.

The neighborhood is characterized by the presence of large workshops for the maintenance of the railways. The location of this neighborhood is central; controlling the roads leading to and coming from the neighborhoods overlooking River Nile and relatively newer neighborhoods such as Nasr City and Heliopolis in East Cairo. Al-Sharabiya is of strategic and economic importance because of the presence of gas refining stations for the "Somid Line" transporting petroleum from El Suez to Alexandria and passing under the River Nile.

Social and Economic Context

An unbalanced social structure is demonstrated in various issues including: Uncontrolled population growth rates, low-income levels, high crime rates, drug abuse, high unemployment rates and deteriorated public services such as health and education. These problems are intensified in three major zones in the neighborhood, which are ranked by the authorities as "unsafe areas"; these zones impose a major safety threat on the rest of the neighborhood. The presence of these three zones as well as the problems discussed above triggered the Egyptian Planning Authorities to initiate this urban development and renewal project.

Demographics

The main data about the neighborhood:

- Total Area: 7.74 km^2
- Inhabited Area: 3.307 km^2
- Population: 249,112 Capita (As of July 1st, 2014)

Site Analysis and Main Landmarks and Development Projects Near the Area

The study area is located southeast of Al-Sharikat Street. Figure 3 shows a detailed site analysis of the study area with the main street network linking the study area with its surrounding neighborhoods. The figure also shows unplanned areas, planned areas, nearby development projects, traffic nodes and railway network.

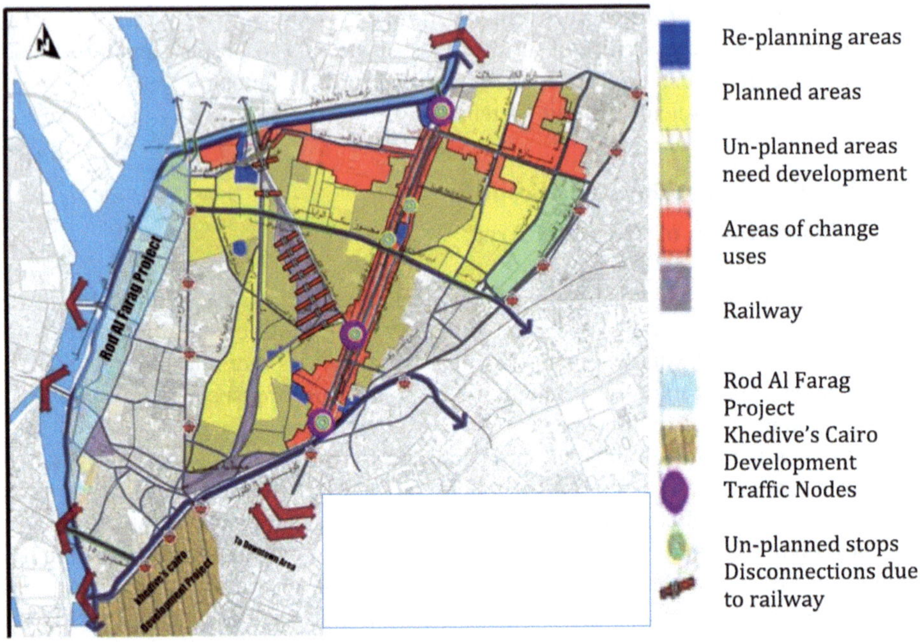

Fig. 3. Site analysis of Al-Sharabiya

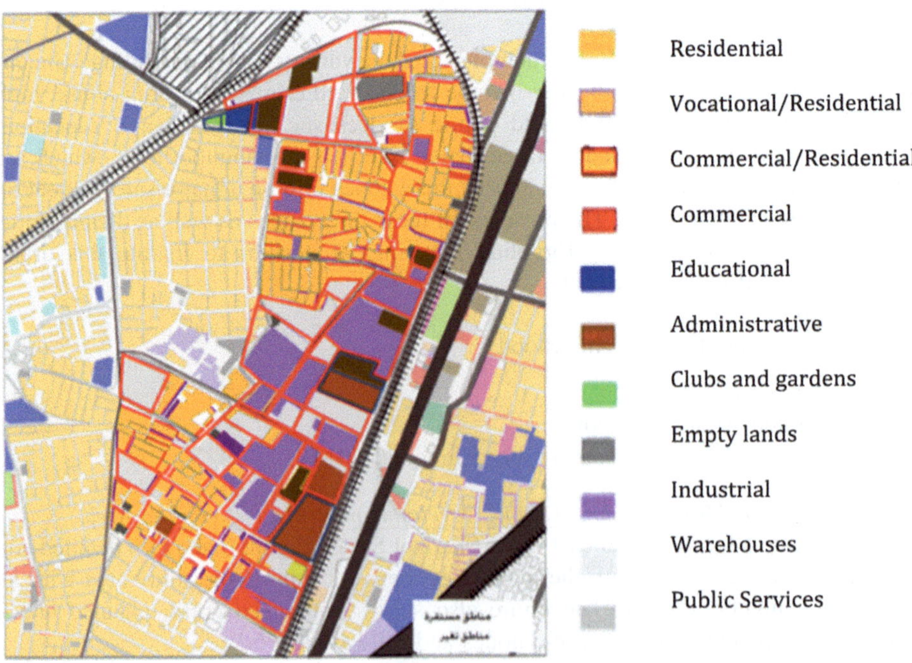

Fig. 4. Existing land uses map

Existing Land Uses
Figure 4 shows the existing land uses in the neighborhood.
Existing Building Conditions
The percentage of buildings with "bad" conditions is 47%; buildings considered as "bad" impose danger on its inhabitants as the structure is unsafe and these buildings are planned to be demolished by the government. The demolition of these buildings will allow the urban planner to widen the streets and propose different land uses for the demolished buildings.
Existing Building Heights
According to the information collected by the urban planners:

- 58% of the buildings are 3 stories or less
- 29% of the buildings are more than 3 and less than 6 stories
- 13% of the buildings are 6 stories and up till 12 stories

Existing Infrastructure Networks

- Transportation Network

The existing road network is shown in Fig. 5; the same figure that shows the existing land uses as mentioned before.

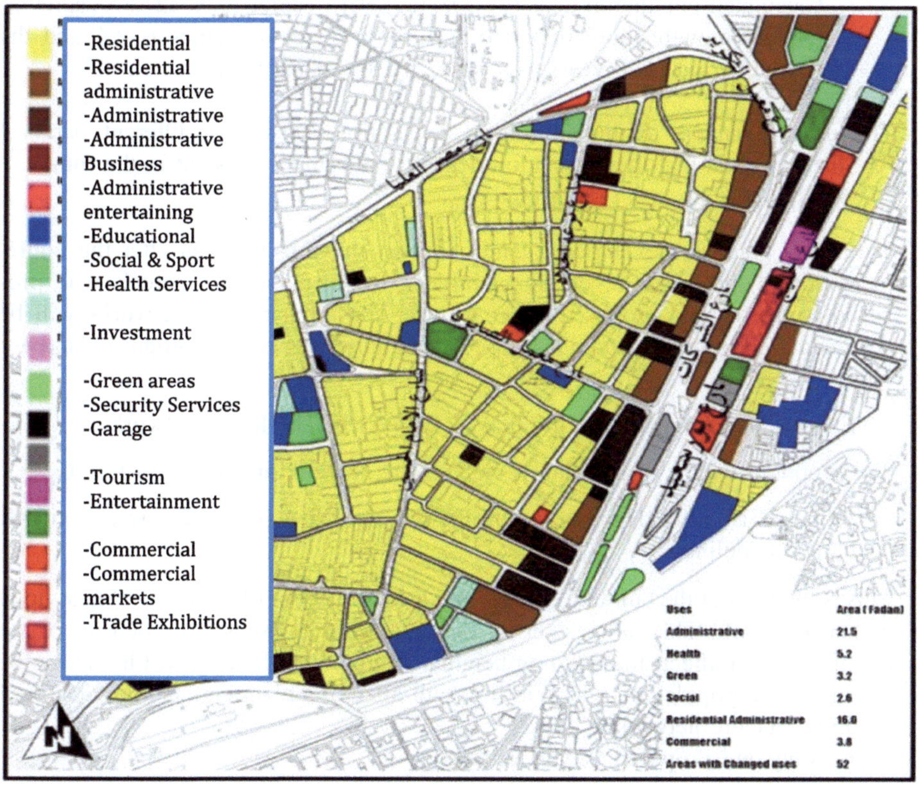

Fig. 5. Proposed land uses map

- Water and Sewer Networks

The location of the existing water and sewer networks was not obtained from the authorities however the location of the main water pipes and sewer lines feeding the neighborhood was located along the main street.

Existing Street Conditions

Not enough information were collected for the street conditions, however based on field visits the majority of the street are not paved with asphalt.

3.2.2 Identification of the Main Influencing Parameters

- *Urban Planning Parameters*
  - Implementation process of the development project

  In order to fit in the infrastructure networks narrow streets are widened; this would involve demolition of the buildings with bad conditions as discussed earlier. People are re-located to live in different areas in areas where construction will take place then they will move back into their buildings.
  - Existing and Expected Future Population Density

    Population = 249112 capita
    Total Area = 7.74 km^2
    Calculated population density = 0.032185013 capita/m^2 = 321.85 capita/ha
    Annual growth population rate according to demographics is 2% and the design period is 20 years as advised by the urban planners.

  The urban planners assumed uniform population density in different parts of Al-Sharabiya and that the population density will remain constant in the future.

  - Existing and Proposed Land Use Map

  The urban planners developed a land use map showing the area after the implementation of the urban development project; the map includes parcels with proposed land uses and proposed street widths. The infrastructure networks will be designed to fit in the proposed land use map. Figure 5 shows the proposed land use map.

  - Existing and Proposed Right Of Way (ROW) In The Road Network

  The proposed widths of the main streets of the network are shown in Fig. 6.

  - Proposed Building Heights According To The New (Floor To Area Ratio Or FAR) Guidelines

  This is the ratio of the building's total floor area to the size of the piece of land upon which it is built. This ratio is set by the authorities to control the density of the population as it affects the building heights. The building heights will influence the density of the population, which in turn will affect the planning, and the design of the infrastructure networks.
  - Existing and Proposed Sewer Network

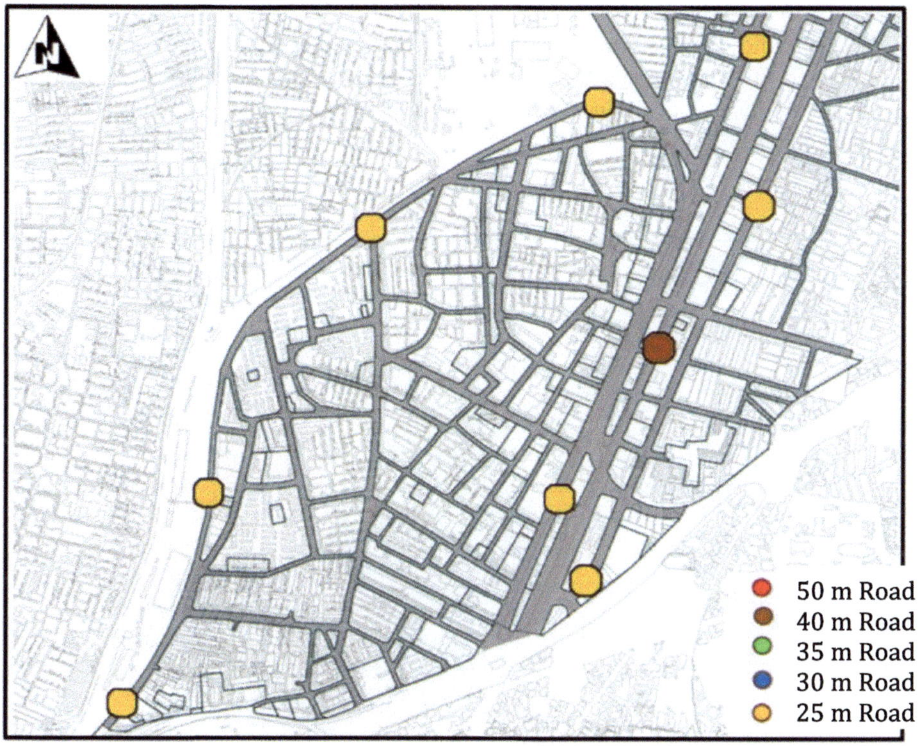

Fig. 6. ROW/proposed street widths

– *Traffic Planning Parameters*

Trip Generation Rates

These rates are the basis upon which traffic planning was conducted, in order to forecast travel demand on each link trip generation rates are assigned to different land uses in each Traffic Analysis Zone (TAZ). These rates were obtained from another traffic study conducted in Borg Al-Arab, which is another low-income neighborhood in Alexandria; we assumed the same trip generation rates for Al-Sharabiya. Table 1 shows the trip generation rates used in the traffic planning process.

Table 1. Trip Generation rates during peak hours

Zone	1	2	3	4	5	Computed P	Given P
1	859	972	1497	1176	1816	6321	6321
2	879	400	1011	931	1179	4400	4400
3	798	596	1504	2272	1754	6923	6923
4	985	862	3570	2085	1610	9112	9112

Mode Choice

The traffic composition was obtained from another traffic analysis study conducted in Al Mattareya Square (About 20 min driving from Al-Sharabiya); the same traffic composition was assumed for our study area, the traffic composition is shown in Fig. 7.

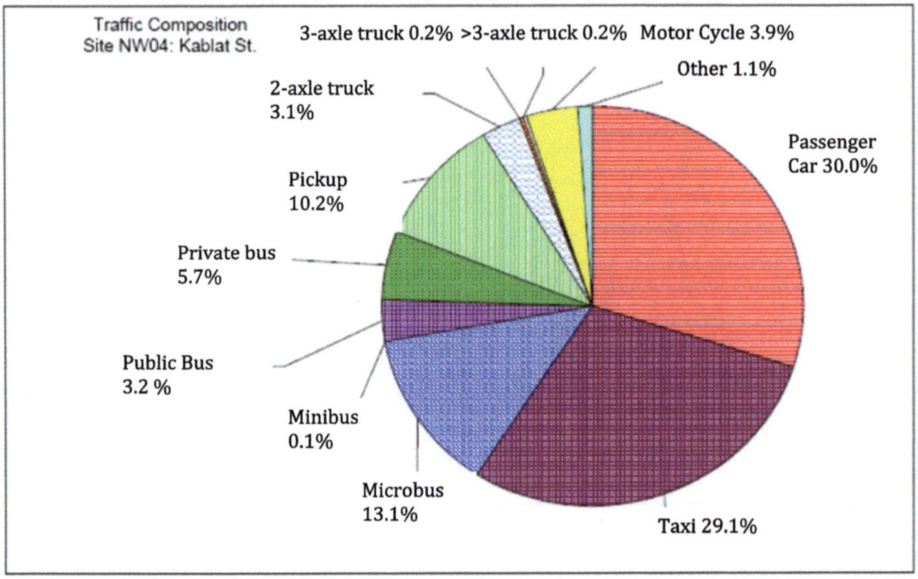

Fig. 7. Traffic composition

Vehicle Occupancy factors

These are the factors that give the number of passengers in each type of vehicles, which allows the conversion of the number of trips into equivalent passenger cars per hour (EPCPH).

Both the engineering parameters and the network design parameters will be explained with each infrastructure network

I. Transportation Network

(1) Traffic Planning

In order to design the road network in El Sharabiya a Travel Demand Study was conducted in order to identify the traffic volumes on each road in the study area.

- *Engineering Parameters*
  - Friction Factor, Fij is inversely proportional to the travel time.
  - Passenger Car Equivalent (PCE) Factors

- *Network Design Parameters*
  A traditional 4-Steps Travel Demand Model was conducted; the process was as following:
  - Trip Generation
    The first step is to delineate the study area boundaries into Traffic Analysis Zones (TAZ) as shown in Fig. 8. The centroids of these zones are marked with red dots. Trip generation rates obtained in Table 1 were used for each TAZ. Several Assumptions were made as following:
      For the Residential land use; 50-apartments/1000 m^2 was assumed, this assumption was made to estimate future demand traffic.
      For the Educational land use; 8.8 m^2 (campus area/student) was assumed.
      For the Health land use; 0.1 beds/m^2 was assumed.

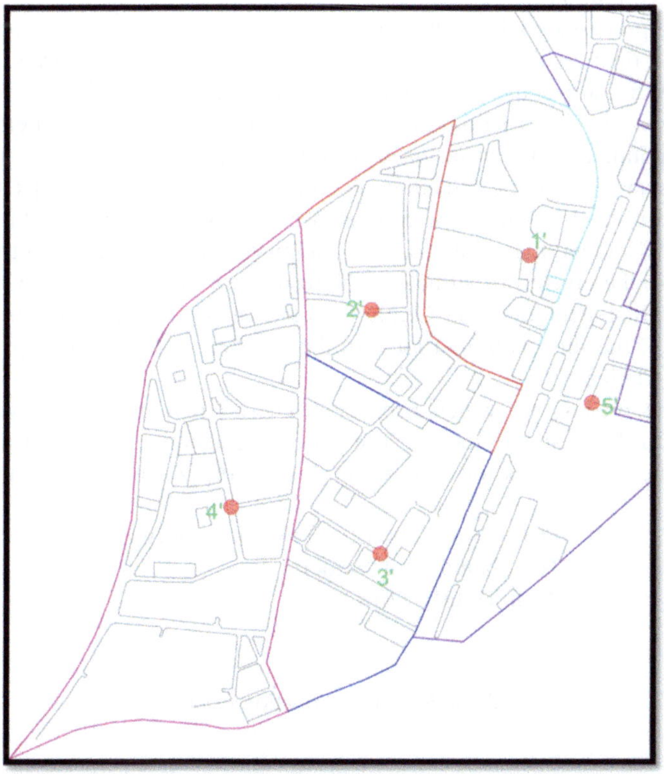

Fig. 8. Traffic Analysis Zones (TAZs)

The results of the trip generation are shown in Table 2.

Table 2. Trip generation results during peak hour PM

		Peak AM Hour			Peak PM Hour		
Land Use	Independent Variable	Total Trips Factor	Attraction	Generation	Total Trips Factor	Attraction	Generation
Residential	Veh/hr	0.44	17%	83%	0.54	67%	33%
Education	Students	0.46	70%	30%	0.15	30%	70%
Adminstrative	Surface area (m2)	0.016	90%	10%	0.016	10%	90%
Commercial	Surface area (m2)	0.052	30%	70%	0.16	50%	50%
Health	Beds	1.8	30%	70%	2	70%	30%
Religious	Surface area (m2)	0.004	20%	80%	0.008	50%	50%
Hotel	Occupied Rooms	0.58	40%	60%	0.78	60%	40%
Entertainment	Surface area (m2)	0.003	50%	50%	0.08	30%	70%
Garage	Surface area (m2)	0.014	30%	70%	0.015	70%	30%

– Trip Distribution

This is the process by which all trips generated in a study area are allocated among zones. The final result of this step is the numbers of zone-zone trips; this will be represented in an origin-destination matrix. The equation most widely used is the Gravity Model. The friction factor was obtained as discussed before and the socio-economic adjustment factor is taken to be equal 1; since all the 5 zones have the same social, economic and cultural conditions. The zone-zone trips iterations are singly constrained in which the computed trips produced equals to the given trips produced as shown; the iterations were computed until the computed trips attracted are equal to the given trips attracted. The results are shown in Table 3, which is the third and final iteration. The Origin-Destination Matrix is shown in Fig. 9; the thickness of the lines is proportional to the number of zone-zone trips.

– Mode Choice

This is the process by which trips between different traffic analysis zones are allocated to the available modes. The following equation was used to obtain the total number of vehicles for each origin-destination (Zone-To-Zone) trips:

$$\begin{aligned} &\Sigma\,(\text{Traffic Composition (i)} * \text{Vehicle Occupancy (i)}) \\ &\quad * \text{Total Number of Vehicles} \\ &\quad = \text{Peak Hourly Trips} \end{aligned}$$

Table 3. Trip distribution results

	Peak Hour PM		
Zones	Trip Production	Trip Attration	Total Trips during Peak Hour PM
1	6321	3526	9847
2	4400	2528	6928
3	6923	6392	13315
4	9112	6133	15245
5	5884	4727	10611
Sum	32640	23306	55946

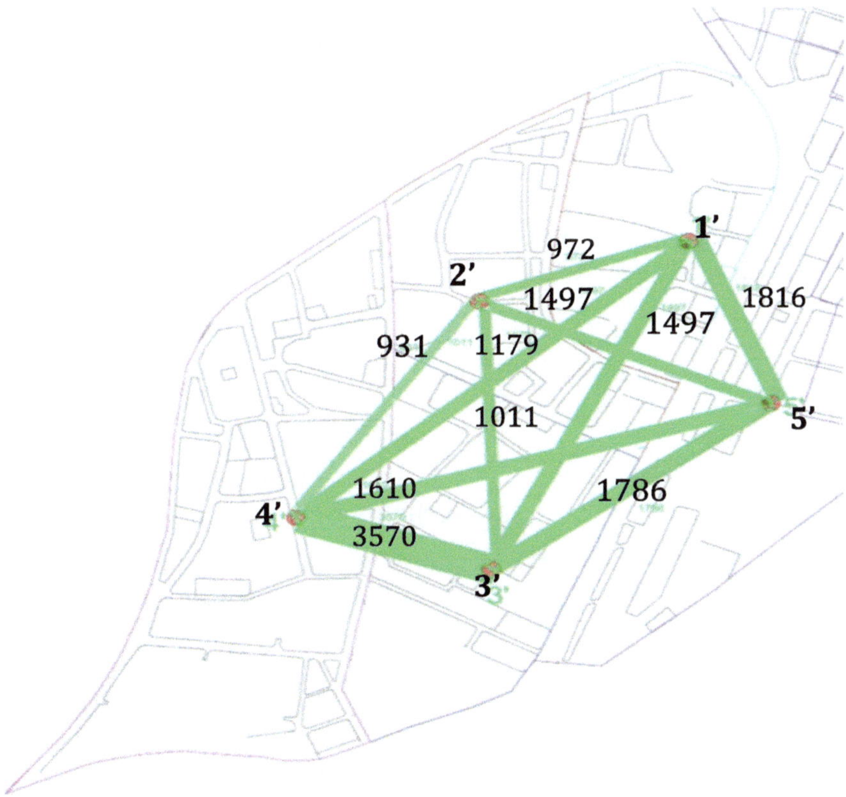

Fig. 9. Origin destination matrix

The traffic composition and the vehicle occupancy were obtained as discussed before. The total number of vehicles was obtained using the equation shown before, and then this traffic was divided again according to the traffic composition.
The number of vehicles was multiplied with passenger car equivalent factors.
Finally the total trips between zones are converted to Equivalent Passenger Car Per Hour (ECPH) during the PM peak hour.

– Traffic Assignment

The process of determining which routes will be used and the number of passenger cars expected on each route during peak hours. Figure 10 shows a links and nodes map for the study area. There are 3 traffic assignment methods that were used, the first is the stochastic traffic assignment in which routes with less travel times are assigned more trips as a probabilistic function; the results for the stochastic traffic assignment between zones 1 and 2 is shown in Table 4. The second method used is system optimization in which the lowest number of total vehicles hours of travel is achieved; the results for the system optimization between zones 1 and 2 are shown in Table 5. The third method used is user equilibrium in which travelers can't improve their travel times by unilaterally changing routes; the results for the user equilibrium assignment between zones 1 and 2 is shown in Table 6.

(2) Cross Section Design

- *Engineering Parameters*
  - The Design Speed For local roads and collectors = 40 km/h and for arterials = 50 km/h

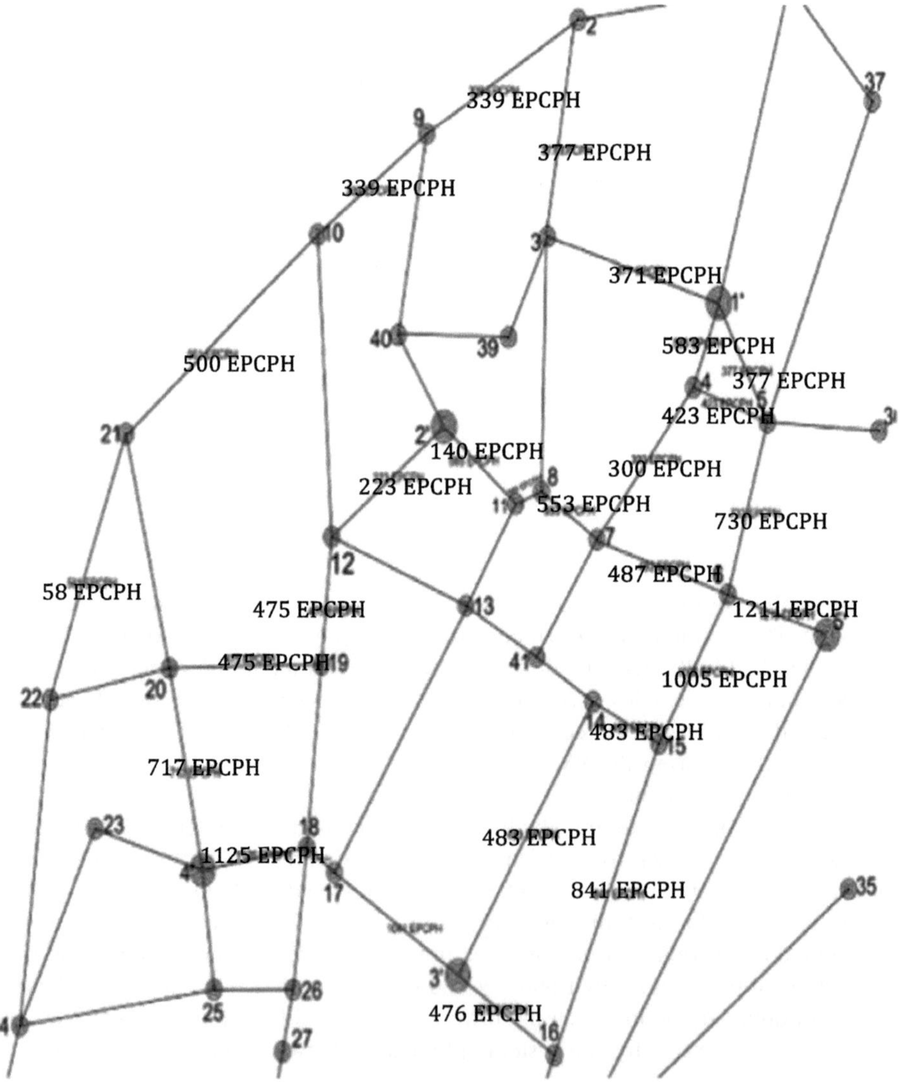

Fig. 10. Links and nodes map

Table 4. Stochastic traffic assignment between Zones 1–2

	Links	Lengths (Km)	Travel Time (min)	Probability	(Trips/Hour)
Route 1	1'-3	0.29832	0.44748		
	3_8	0.33885	0.508275		
	8_11	0.04466	0.06699		
	11_2'	0.26879	0.403185		
			1.42593	0.251372172	245
Route 2	1'-4	0.13311	0.199665		
	4_7	0.26103	0.391545		
	7_8	0.10644	0.15966		
	8_11	0.04466	0.06699		
	11-2'	0.26879	0.403185		
			1.221045	0.308530158	300
Route 3	1'-4	0.13311	0.199665		
	4_5	0.14856	0.22284		
	5_6	0.27349	0.328188		
	6_7	0.19949	0.299235		
	7_8	0.10644	0.15966		
	8_11	0.04466	0.06699		
	11_2'	0.26879	0.403185		
			1.679763	0.195019898	190
Route 4	1'-3	0.29832	0.44748		
	3_2	0.1713	0.25695		
	2_9	0.15142	0.181704		
	9_40	0.28561	0.428415		
	40_2'	0.09116	0.13674		
			1.451289	0.245077772	239

Table 5. System optimization traffic assignment between zones 1–2

	Links	Lengths(Km)	Velocities (km/hour)	Free Flow Travel Time(min)	Free Flow Travel Time(hours)	Volumes (Trip/hour)	Travel Time (hour)	Travel Time(mins)
Route 1	1'-3	0.29832	40	0.44748				
	3_8	0.33885	40	0.508275				
	8_11	0.04466	40	0.06699				
	11_2'	0.26879	40	0.403185				
				1.42593	0.0237655	317	0.028197894	1.691873623
Route 2	1'-4	0.13311	40	0.199665				
	4_7	0.26103	40	0.391545				
	7_8	0.10644	40	0.15966				
	8_11	0.04466	40	0.06699				
	11-2'	0.26879	40	0.403185				
				1.221045	0.02035075	341	0.025466083	1.527964974
Route 3	1'-3	0.29832	40	0.44748				
	3_2	0.1713	40	0.25695				
	2_9	0.15142	50	0.181704				
	9_40	0.28561	40	0.428415				
	40_2'	0.09116	40	0.13674				
				1.451289	0.02418815	314	0.028536012	1.712160704

Table 6. User equilibrium traffic assignment between zones 1–2

	Links	Lengths(Km)	Velocities (km/hour)	Free Flow Travel Time(min)	Free Flow Travel Time(hours)	Volumes (Trip/hour)	Travel Time (hour)
Route 1	1'-3	0.29832	40	0.44748			
	3_8	0.33885	40	0.508275			
	8_11	0.04466	40	0.06699			
	11_2'	0.26879	40	0.403185			
				1.42593	0.0237655	305	0.0275751
Route 2	1'-4	0.13311	40	0.199665			
	4_7	0.26103	40	0.391545			
	7_8	0.10644	40	0.15966			
	8_11	0.04466	40	0.06699			
	11-2'	0.26879	40	0.403185			
				1.221045	0.02035075	372	0.0275751
Route 3	1'-3	0.29832	40	0.44748			
	3_2	0.1713	40	0.25695			
	2_9	0.15142	50	0.181704			
	9_40	0.28561	40	0.428415			
	40_2'	0.09116	40	0.13674			
				1.451289	0.02418815	295	0.027575111

- Capacity for roads was taken to be 300 veh/hr/lane
- The designed level of Service (LOS) is C. The corresponding (V/C) was assumed to be 0.8 according to the Highway Capacity Manual.
- Service Volumes; based upon the results of the traffic assignment discussed before, traffic was assigned on each link from each zone-zone trip distribution.
- Adjustment Factors for multilane highways were not used because they cannot be applied to this case study.

* *Network Design Parameters*
  - Number of Lanes

    The number of lanes was calculated according to the following equation:

$$\text{Number of Lanes} = \text{Service Volume}/(\text{Capacity} * (\text{V/C}))$$

The results for the number of lanes for some of the links are shown in Table 7.

Table 7. Number of lanes for some of the links

Road	S.V (Vph)	V/C LOS C	Number of Lanes	Approximated Number of Lanes
1'-4	583	0.8	2.43	3
4_7	300	0.8	1.25	2
7_8	653	0.8	2.72	3
8_11	585	0.8	2.44	3
11-2'	585	0.8	2.44	3
1'_5	377	0.8	1.57	2

- Lane Width varied in different cross section design from 2.5–3.5 m. This was done to consider the Right Of Way discussed before and to accommodate the number of lanes, the median width, parking lane width and the pedestrian walkway width.

- Median Width was taken to be 2 m
- In order to provide parking for cars without disturbing the traffic flow, a parking lane was added on each direction, its width was taken to be 2.5 m
- Cross Slope was taken to be 1.5%.

Cross-Section design sample is shown in Fig. 11

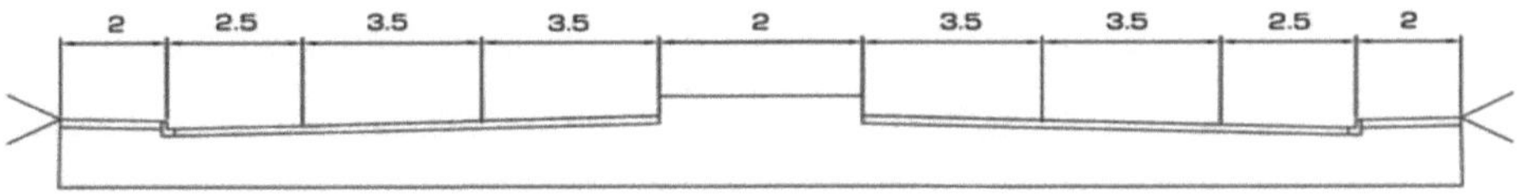

Fig. 11. Cross-section sample

(3) Horizontal Alignments

- *Engineering Parameters include:*
  - PIEV Time; as recommended by the AASHTO 2.5 s is the design value.
  - Minimum safe radius/maximum safe curvature

    For design speed of 40 km/h, the maximum factor of side friction (Fs) = 0.17

    Superelevation = 0.10

    R min = 402/9.81(0.10 + 0.17) = 12 m

    D max = 1746/709 = 2.4
  - Length for Superelevation Runoff

    Take Superelevation runoff = 110 m

    For 4-lane, increase values by 50%; Superelevation runoff = 165 m
- *Network Design Parameters* include Point of Curvature, Point of Tangency and Point of Intersection. The network was designed using Civil 3D software.

(4) Vertical Alignments and Drainage Network

- *Engineering Parameters*
  - Maximum and Minimum Grades
  - Stopping/Passing Site Distances
  - Length of each vertical curve was checked that it is larger than both the SSD and PSD; according to the AASHTO tables.
- *Network Design Parameters*
  - Vertical Point of Curvature
  - Vertical Point of Tangency
  - Vertical Point of Intersection
  - Vertical Profiles were designed using Civil 3D software, design sample of profile is shown in Fig. 12.
- Catch Basin Map
  - After the vertical profiles are designed for the network, points at lower elevation where surface water from rain converges to are identified and catch basin system is installed at these points

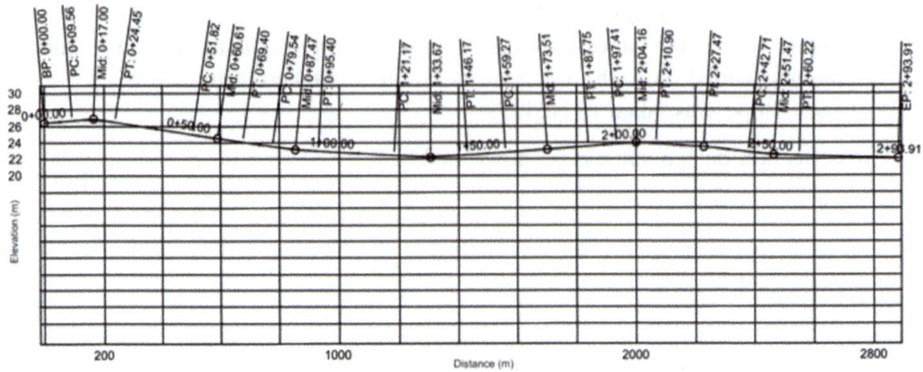

Fig. 12. Vertical profile sample

- The purpose of the Drainage Map is to know where water will be drained in every part of the studied area so no accumulation occurs, this is important for the life time of the asphalt pavement and for traffic congestion
- Figure 13 shows the drainage map showing the flow of water

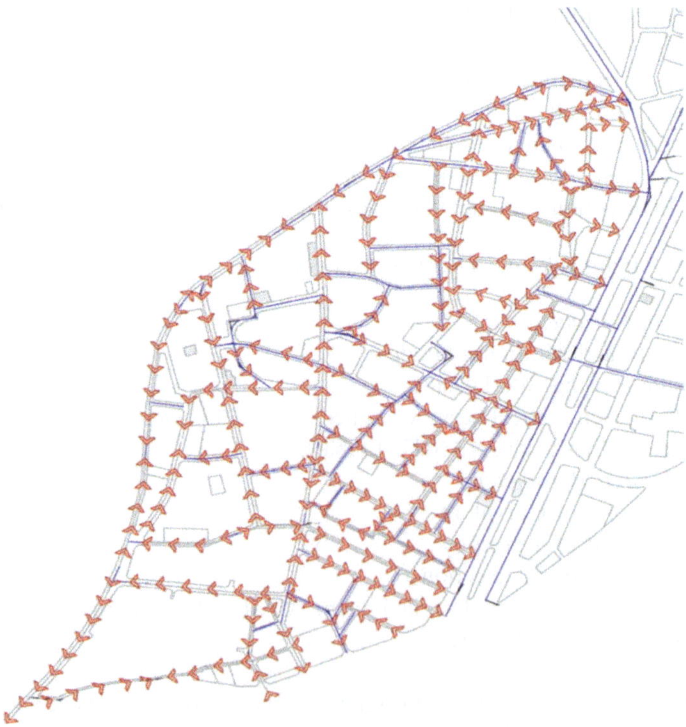

Fig. 13. Drainage map

– Figure 14 shows the catch basin map with the red dots identifying the points that need catch basins to be installed.

Fig. 14. Catch basin map

(5) Intersection Design

In this case study, the intersections designed were all at-grade intersections.

- *Engineering Parameters*
  – Minimum Radii for intersection curves
  – According to AASHTO design tables, the suggested minimum radius for design speed of 25 mph or 40 km/h is 45.72 m or 150 ft.
- *Network Design Parameters*

In this case study, the intersections designed were all at-grade intersections; flared intersections were commonly used. Intersections were designed using Civil 3D software

(6) Asphalt Pavement Design

- *Network Design Parameters*

  These include: Surface course (bitumen), base course and sub base course thicknesses.

- For main roads (with heavy trucks); Surface (Bitumen) course thickness is 5 cm, base course thickness is 20 cm and the sub base thickness is 25 cm
- For secondary roads (in which heavy trucks are not allowed); Surface (Bitumen) course thickness is 4 cm, base course thickness is 10 cm and the sub base thickness is 15 cm

II. Water Distribution Network

- *Engineering Parameters*
  - Average water consumption rates

  The Egyptian Code assigns values for the average water consumption for each land use. These values are then multiplied by the independent variable for each land use. A network of large loops is then designed as shown in Fig. 15. The total average water consumption in each loop (design area) is calculated by summing up all the values of average water consumption from each designed block.

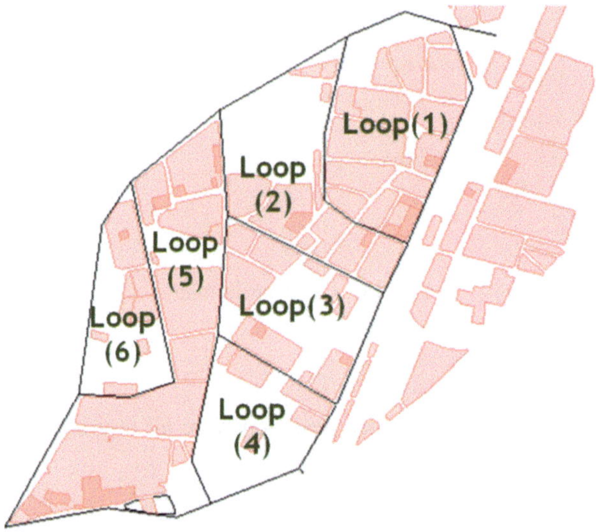

Fig. 15. Layout of the entire water network composed of large loops

  - Peak Hourly Factor

  According to code, the peak hourly factor is taken to be 2.5. This factor is multiplied by the average water consumption in order to obtain the peak hourly demand, which will be assigned to the junctions and used in the design process. The peak hourly demand of the loops is shown in Table 8.
  - Junction elevations

  In order to simulate the designed water network using WaterCad software, elevations are obtained from the contour map of the site and assigned to the junctions.
- *Network Design Parameters*
  - Elevated Tanks

Table 8. Peak hourly demands of the water network

Loop	Peak Hourly Demand(L/day)
1	4124775.334
2	2686353.484
3	2590772.444
4	2871426.366
5	5006968.382
6	1508739.101

In order to simulate the main water pipe feeding the networks using WaterCad, two elevated tanks were defined on the software. The elevations of the tanks, as shown in Table 9, model the energy heads at two points (inlets) in the main water pipe. The head loss between the two points was calculated using Hazen Williams equation for frictional head loss and head loss equation.

Table 9. Elevated tanks data

Label	Elevat ion (Base) (m)	Elevation (Minimum) (m)	Elevatio n (Initial) (m)	Elevation (Maximum) (m)	Volume (Inactive) (MG)	Diam eter (m)	Flow (Out net) (m /day)	Hydrau lic Grade (m)
T-1	28.5	52.5	57.5	62.5	0	10	12,657	57.5
T-2	27	48	53	58	0	10	6,133	53

– Pipe Materials

Martial Used: uPVC due to its availability in the region, as well as having diameters that are less than 500 mm. This material is economic and it satisfies our main goal in this project, which is developing the neighborhood at the lowest cost possible.

Hazen Williams Coefficient = 140

– Pipe Diameters

The pipe diameters ranged between 150 and 300 mm.

– Flow Velocity

The flow velocity ranged between 0.33 and 1.55 m/s.

A sample of the pipes results for the large network of loops is shown in Table 10.

Table 10. Sample of pipe results

Label	Length (Scaled) (m)	Start Node	Stop Node	Diameter (mm)	Flow (m^3/day)	Velocity (m/s)	Headloss Gradient (m/m)
P-1	142	J-236	J-145	150	-384	0.25	0
P-2	99	J-145	J-111	150	-349	0.23	0
P-6	62	J-161	J-162	200	3,275	1.21	0.006
P-7	104	J-162	J-166	150	1,211	0.79	0.004
P-9	99	J-161	J-169	250	6,476	1.53	0.007
P-10	48	J-169	J-166	150	722	0.47	0.001
P-11	138	J-169	J-172	250	5,663	1.34	0.005
P-12	41	J-172	J-173	150	1,528	1	0.006
P-13	39	J-172	J-178	250	4,064	0.96	0.003
P-14	42	J-173	J-174	150	1,482	0.97	0.006
P-15	105	J-174	J-175	150	1,436	0.94	0.005
P-16	31	J-175	J-186	150	844	0.55	0.002
P-17	176	J-166	J-176	150	1,839	1.2	0.008
P-18	97	J-162	J-279	150	1,998	1.31	0.01
P-19	120	J-279	J-188	150	1,941	1.27	0.009
P-20	115	J-188	J-187	150	-251	0.16	0
P-22	45	J-187	J-280	150	-285	0.19	0
P-23	34	J-280	J-186	150	1,437	0.94	0.005
P-24	19	J-176	J-280	150	1,755	1.15	0.008
P-26	121	J-186	J-190	150	2,159	1.41	0.011

– Pressure in Junctions

The pressure in junctions ranged between 21 and 30 m H_2O; this is sufficient to pump water to (5–7) stories buildings without the use of pumps.

III. Sewer Network

In this case study a combined sewer system is designed, which is a system that collects both sanitary sewage and storm water runoff. This system was selected because the study area is not considered rainy but rather arid, which makes this option more economic rather than separate sewer system for sewage and storm water. Moreover, this system will collect storm water drained from the drainage system discussed before.

- *Engineering Parameters*

  – Average Wastewater consumption rates

  Q_{av} (Wastewater) = 0.8−0.9 Q_{av} (Water Consumption). To be more conservative in our design we multiplied by 0.9.

  – Minimum Allowable Slopes of sewer lines

  The Egyptian code of practice specifies values of minimum allowable slopes for each pipe diameter.

- Minimum and Maximum velocity

  Minimum velocity "Self-Cleansing Velocity" is about 0.6 m/s and can be as low as 0.45 m/s; to prevent or reduce excessive deposition of solid material in the sewers.

  Maximum velocity is about 3 m/s and can reach 4.5 m/s; to prevent corrosion, scour and other hydraulic problems.

  According to the ECP, the most optimum range of velocities is between 0.6 m/s and 1.5 m/s.

- Minimum depth of sewer lines

  Sewers must be placed at a depth such that they have sufficient cushioning to prevent breakage due to ground surface loading; minimum cover depths must be specified according to the ECP as following:

  Minimum of 1 m depth below ground surface and top of the pipe for sewer lines passing under roads with traffic.

  Minimum of 0.6 m depth below ground surface and top of the pipe for sewer lines passing under pedestrian pathways.

- *Network Design Parameters*

  The sewer system is designed as a dendritic network converging in the downstream direction without closed loops. The sewer system is designed for "Gravity Flow".
  - Conduit material selected is uPVC; the manning coefficient for the uPVC is **0.012**
  - Conduits Diameters
  - Conduits Slopes
  - Crown or invert elevations for each pipe in the system
  - Flow Velocity

  Table 11 shows some of the conduits diameters, slopes, invert elevations and flow velocity.

Table 11. Sample of conduits results

Label	Invert (Start) (m	Invert (Stop) (	Length ((m)	Slope (m/m)	Diameter (mm)	Flow (m^3/day)	Velocity (m/s)	Depth (Middle) (m)	Capacity (Full Flow) (m^3/day)
CO-8	19	18.5	120.5	0.004	400	3,253.31	0.59	0.2	12,554.07
CO-13	27	26.5	142.7	0.004	200	732.35	0.35	0.15	1,817.44
CO-14	27.5	26	122.6	0.012	200	1,197.12	0.64	0.13	3,395.68
CO-16	26.5	26	142.2	0.004	200	918.99	0.4	0.16	1,820.31
CO-19	24.25	22.75	55.4	0.027	200	602.01	0.57	0.08	5,049.30
CO-20	22.75	22.5	97	0.003	200	839.39	0.35	0.17	1,558.46
CO-23	20	19.5	79.6	0.006	400	3,861.99	0.72	0.2	15,448.52
CO-24	23	20	131.1	0.023	200	231.28	0.31	0.06	4,643.60

- Drop Manholes

When sewer lines join in a deep Sewer and a vertical drop of more than 0.9 m is required, a drop manhole is added to maintain the normal slope of the sewer pipelines in case the ground slope is excessive; this prevents supercritical to subcritical flow transition. Figure 16 shows a conduit profile where a drop manhole is installed to maintain the normal slope of the sewer pipeline

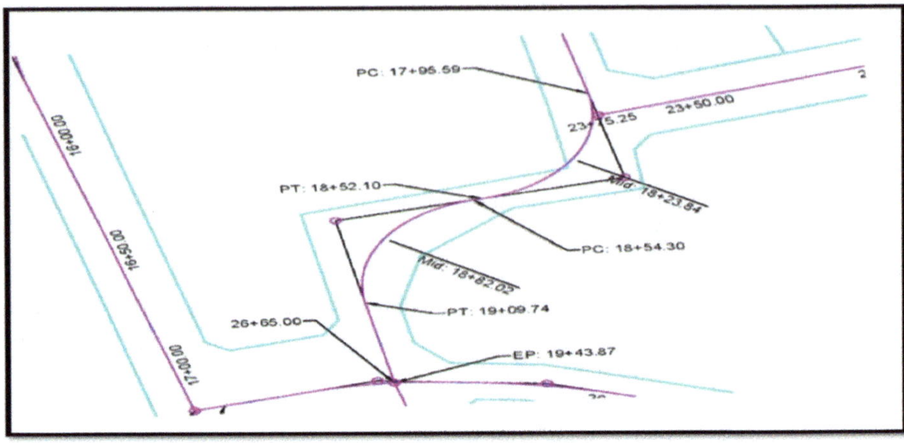

Fig. 16. Node 1 problem

– Flushing System

Flushing Tanks are installed under road surfaces at the beginning of sewer lines with velocities lower than 0.45 m/s. This is done to prevent settlement of organic matter. These tanks are connected with clean water pipes. As the tank is filled with water, the syphon or the flushing valve automatically flushes water in the Sewer pipes increasing the velocity. This automatic flushing system is mentioned in the ECP so it is allowable, however the common practice in Egypt involves cars equipped with pressurized water pipes. Occasionally, these cars come and connect the water pipes to the manholes at the beginning of the sewer lines flushing down any settled matter.

3.3 Collection of Data

Sources from which data are collected include the following:

- Site visits and field work
- Online scholarly articles and studies
- Central Agency for Public Mobilization and Statistics, Heliopolis, Cairo.

Missing data included:

- Actual Traffic Counts for the study area
- Analysis of the streets conditions
- Information regarding traffic and pedestrian accidents
- Existing water network
- Existing sewer network
- Location of nearby sewage treatment plant

Unfortunately, due to the time constraints these data was not obtained.

3.4 Modeling the Existing Networks Using Appropriate Software

Since no sufficient data regarding the existing networks was obtained, and because the urban development project proposed different land uses with a new road network; designing new networks and modeling their performance was more appropriate.

3.5 Analyzing the Performance of the Existing Networks

The performance of the existing networks was not analyzed as well due to the same reason mentioned in the previous step.

3.6 Making Possible Improvement Suggestions

Suggestions were not made to existing networks since they were neither modeled nor analyzed; however in this case study suggestions were made to improve the network design; this was mentioned in the network design parameters for each network. Moreover, some aspects required suggesting different approaches of design to fit the context of the project, which was mentioned in the network design parameters of each network. Suggestions were also made to the proposed urban development map after finalizing the horizontal alignments, cross section and the intersection design. These suggestions are reported to the urban planning authorities till an agreement is reached. Examples of the improvement suggestions include:

1. *Traffic Planning*
   Passenger Car Equivalent Factors were obtained from standard values however the factors for the minibus and microbus are inflated and assumed to be 1.4 and 1.5 respectively to account for misbehavior in driving and shifting between lanes.
2. *Cross Sections*
   The Capacity of the roads was assumed to be 300 passenger car/hour/lane. The capacity was assumed low because of many obstacles on the roads, misbehavior of the drivers, pedestrians crossing at random points rather than through designated areas, poor asphalt pavement quality and sellers occupying parts of the streets.
3. *Horizontal Alignments*
   Node 1:
   - Problem: The minimum radius clashed with the corner of a block on the proposed land use map as shown in Fig. 17
   - Solution: Recommendation to the urban planner to change the block shape as shown in Fig. 18

   Node 2:
   - Problem: Undefined road between two small pieces of land (small gardens) as shown in Fig. 19
   - Solution: Combining both pieces of land to define one road and avoid any confusions to the drivers as shown in Fig. 20

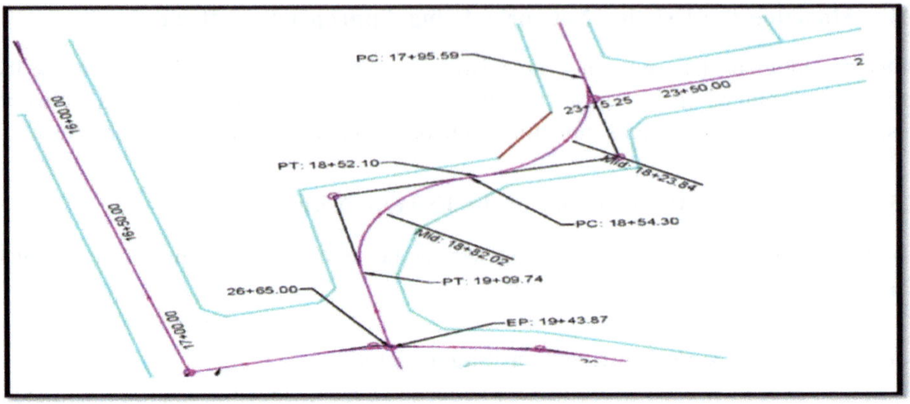

Fig. 17. Node 1 solution

Fig. 18. Node 2 problem

4. *Vertical Alignments*
   Regarding the Maximum Grade; according to AASHTO passenger cars can climb a grade of less than 7% without significant reduction in speed; however in our case for urban slums in Egypt we assumed that passenger cars can climb 5% only without significant reduction in speed and this is due to lower performance of cars in such areas. On the other hand the minimum grade was taken to be 0.25% for drainage since this is an arid area.

Fig. 19. Node 2 solution

Fig. 20. Proposed solution for the challenging intersection

5. *Intersection Design*
   A challenging intersection where the blocks surrounding the intersection don't give a defined shape for the traffic circulation; it was proposed to place a small green block between the blocks to define the circulation as shown in Fig. 21.

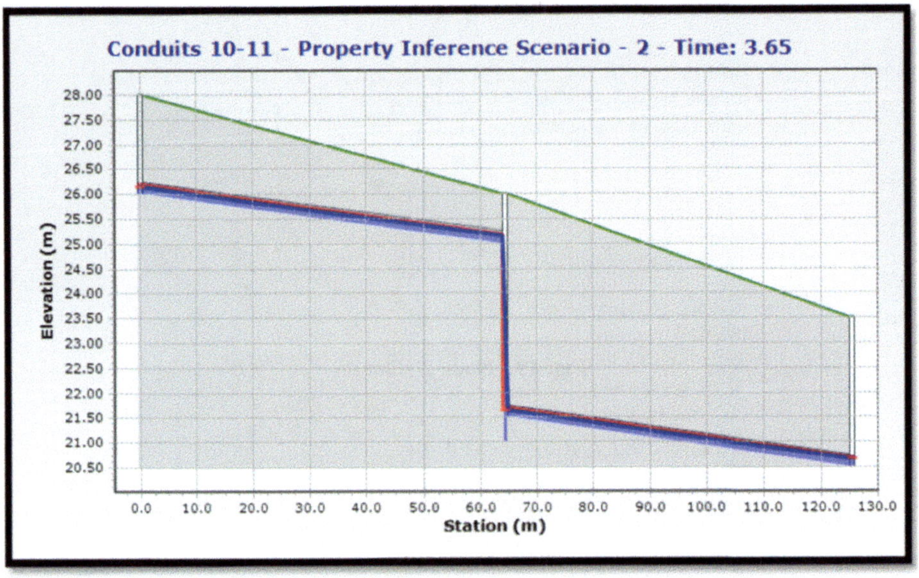

Fig. 21. Conduit profile with a drop manhole

6. *Sewer Network*
   Several iterations of the layout of the sewer network were made; however the major challenge was the presence of a low elevation point between two very high elevation points, the profile of sewer lines connected to these points is shown in Fig. 22. If we are to connected these conduits, sloping down, to the entire sewer network this would mean very low invert levels, unpractical excavation levels extending to more than 10 m depth below the ground surface.

- Solution 1:
  Connecting this sewer line directly to the main sewer line to avoid lowering the invert levels of the network and localize the areas where large excavations are needed; this is shown in a zoomed in picture of the connection is shown in Fig. 23. The profile of the conduits connecting this point to the main sewer line is shown in Fig. 24; it indicates that a great amount of soil has to be excavated.
- Solution 2:
  Addition of a small pump at this low point, giving energy (head) to the sewage to be lifted to a high invert level, this will maintain the invert levels of the network; as shown in Fig. 25.

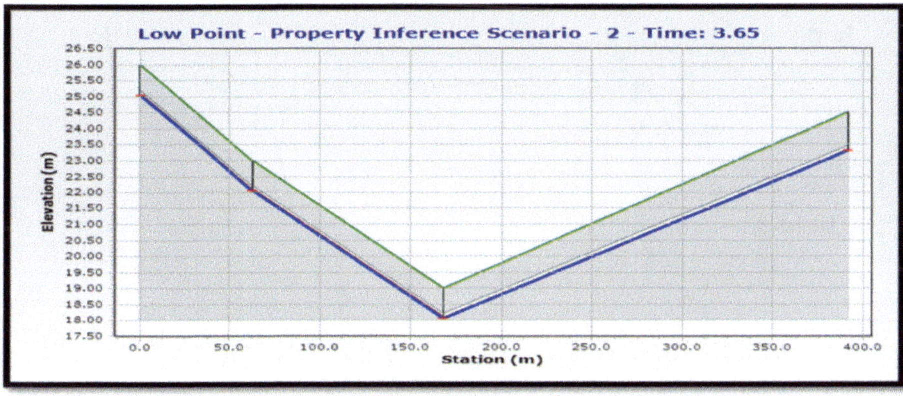

Fig. 22. A zoomed in view of the connection in solution 1

Fig. 23. A zoomed in view of the connection in solution 1

Fig. 24. Profile of the conduits connecting the low point in solution 1

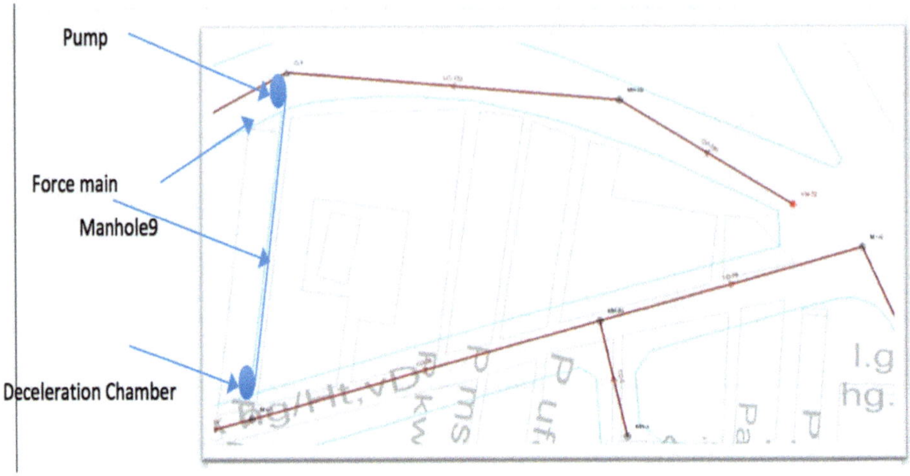

Fig. 25. Solution 2 for the sewer network

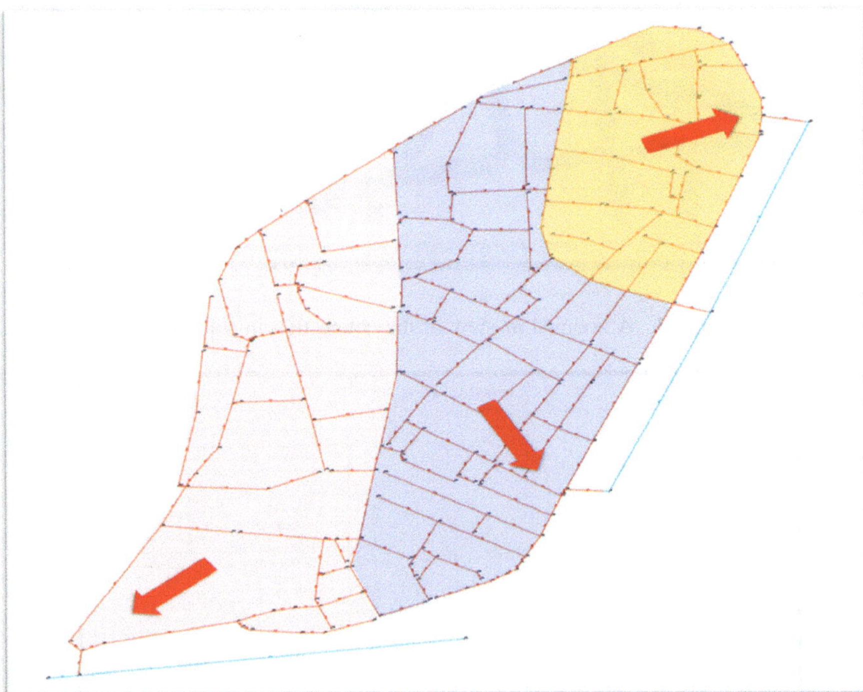

Fig. 26. Final layout of the sewer network and the discharge for each area

The pump is installed at the beginning of the street connected to a force main which must be connected to a deceleration chamber before joining the gravity sewer line to lower down the pressure before joining the gravity sewer line, to avoid turbulent flow. Instead of designing the force main, a manhole was added in the gravity sewer line with an invert level of 22.5 m as if it is equivalent to the energy head provided by the force main in the actual conditions. After several iterations on SewerGems a final layout of the sewer network had been reached and the flow of sewage from each area in the network is shown in Fig. 26.

3.7 Modeling the Performance of the Improvement Suggestions

The suggested network designs were modeled as following:

- Traffic Planning was modeled manually rather than using PTV Visum as the software was not accessible
- Horizontal, vertical and intersection designs were modeled using Civil 3D software
- Water network was modeled using WaterCad software
- Sewer network was modeled using SewerGems software

3.8 Quantification of the Results

After final network designs have been reached and their performance is modeled and accepted, the results are quantified; tables and figures validate the improvement suggestions to different network designs.

4 Conclusion

The proposed solution model for infrastructure development of urban slums has been validated as it has been applied to Al-Sharabiya, which is one of the urban slums in Cairo. Although the application deviated from the proposed model as existing infrastructure networks were not modeled and basically new infrastructure networks were designed; however the new infrastructure networks were not designed as per international standards but the design considered the social and economic context of Egyptian urban slums. Instead of suggesting improvements to the existing networks as proposed by the model, the application of the case study showed that even if new networks are to be designed the suggested improvements can be made to international standards of design. To sum up, the proposed solution model was validated as a methodology for developing infrastructure networks (Transportation, Water and Sewer Networks) in urban slums in Egypt.

Acknowledgments. This paper could have never been completed without the guidance and mentorship of our precious professors in the American University in Cairo: Dr. Safwan Khedr, Dr. Maram El Saudy, Dr. Emad Emam and Dr. Ahmed El Gendy. We would like also to acknowledge the efforts of Eng. Ahmed El Dewainy from the Egyptian Urban Planning authorities who provided us with the urban development project and most of the data we used in order to conduct our study in Al-Sharabiya.

Multi-criteria Analysis for Spatial Modelling of Forest Protection Zones

Ather Ashraf(✉) and Mehwish Arif

Punjab University College of Information Technology, University of the Punjab, Old Campus, Lahore 54000, Pakistan
atherashraf@gmail.com, mehwish.gis@gmail.com

Abstract. Irrational utilization of land and fast developing of infrastructure due to increase in the population has led to the stagnation of the forest landscape. Deforestation results in, the demise of forest cover for the purpose of agriculture development or implementing urban architecture or it could be an implication of unrestricted grazing. This documentation is focused to represents and consul a procedure and process for the forest retention and conservation. Spatial modelling has been employed to contemplate the land cover changes using Markov chain and CA-Markov to predict LULC for 2020. Deforestation has become a leading environmental subject due to its effects and which plays a crucial part in destruction of natural ecology. Therefore, defining the fitness criteria of forest and finding the effective parameters causing deforestation can help in forest management and conservation. For the assessment of Forest fitness criteria Slope, EVI, Forest Land use type and distance to roads, drainage network, and settlement have been used as criteria of factor maps. Distance importance chart has been developed for each factor based on the threshold values with respect to forest. By using Multi Criteria Analysis (MCA) process, the Forest Protection Zones have been pointed, the pockets that are most susceptible to deforestation in 2020 have been identified and forests have also been mapped that are most important to protect in future. Thematic land use land cover types of 1992, 2005 and 2009 of Landsat sensor have been classified. The spatial extent decrease from 1999 to 2009 of forest has been observed to be 9.76% and if the proper measures will not be taken then from 2009 to 2020 the predicted forest spatial extent decrease has been observed to be 4.59%. The projected Land use Land Cover map for 2020 would provide useful inputs to decision makers. The forests are located at lower elevation and close to roads have been observed most vulnerable to clearance. Distance to roads and settlements are explored as the major factors for loss of forest.

Keywords: Cellular automata · Markov chain · Deforestation · Multi criteria analysis · Land use land cover

1 Introduction

Forests are carbon sinks, prevents soil erosion acts as a windbreak and supports biodiversity. In many areas of rural societies the forest trees and shrubs are sources of many wood and non-wood products. Forests contribute to soil and water conservation.

J. Calautit et al. (eds.), *Towards Sustainable Cities in Asia and the Middle East*,
Sustainable Civil Infrastructures, DOI 10.1007/978-3-319-61645-2_14

Forest animals play important role in forest ecology. Sustainable Forest management means increase in medicine, food, raw materials and employment opportunities. Forest management is important for environment such as carbon sequestration and also for the conservation of water, soil and biodiversity. It also supports in preventing deforestation and forest degradation. FAO defines Forests as "Land spanning more than 0.5 ha with trees higher than 5 m and a canopy cover of more than 10%, or trees able to reach these thresholds in situ. It does not include land that is predominantly under agricultural or urban land use" (Wilkie 2010). Forests are designated for different purposes i.e. protected areas, production, Protection of soil and water, Conservation of Biodiversity and social services (Wilkie 2010). According to FAO, the conservation of forest to another land use or below the 10% threshold the reduction of tree cover is deforestation (Anonymous 2000). Deforestation and forest degradation are important issues with respect to climate change. Forest resources are decreasing very fast in Pakistan. Most of the Forests land in Pakistan has suffered from encroachments. Deforestation is also the result of uncontrolled grazing and rapid unchecked cutting of trees. The rate of deforestation is alarming as all the forest area will convert to other land use types in coming years.

In January 1994, the IUCN General Assembly meeting in Buenos Aires approved the new system. Guidelines were published by IUCN and the World Conservation Monitoring Centre later that year (IUCN 1994). These set out a definition of a "protected area" – "An area of land and/or sea especially dedicated to the protection and maintenance of biological diversity, and of natural and associated cultural resources, and managed through legal or other effective means – and six categories:

Areas managed mainly for: I. Strict protection [(a) Strict nature reserve and (b) Wilderness area], II. Ecosystem conservation and protection (i.e., National park), III. Of natural features (i.e., Natural monument), IV. Conservation through active management (i.e., Habitat/species management area), V. Landscape/seascape conservation and recreation (i.e., Protected landscape/seascape), VI. Sustainable use of natural resources (i.e., Managed resource protected area)" (Dudley 2008). Protected areas must be established for the conservation of biodiversity. They are the key stones to maintain ecological processes and only hope for threatened species from extinction. Protected areas provide us the opportunities to interact with nature. They provide us natural space that is deficient in an increasingly managed and crowded planet. Protected areas also symbolize a guarantee to future generations. The Convention on Biological Diversity (CBD) was agreed at the 1992 Earth Summit and, to date, has been approved by 188 nations. It has three major goals are: the conservation of biological diversity; the sustainable use of its components; and the equitable sharing of benefits arising out of the utilization of genetic resources (Dudley and Parish 2006). The Program of Work on Protected Areas identifies four program elements, 16 goals and 92 associated activities for state parties. However, although the list of expected outputs is long, they all relate back to the central objective, encapsulated in goal 1.1 as: "To establish and strengthen national and regional systems of protected areas integrated into a global network as a contribution to globally agreed goals". The CBD highlights that the purpose is not only to increase the number of protected areas but also should be designed and located in the best places to conserve biodiversity and that they should be determined by a multi-stakeholder process (Dudley and Parish 2006).

The most important point is the selection of Forest protected areas. The process of selection begins with the IUCN definition of protected areas and then further refined by reference to the IUCN categories. If the area meet the IUCN definition of a protected area then assign to one of the IUCN category else it will not be considered as a forest protected area. The next process is to calculate proportion of forest in the protected area. Remove any area of trees that do not meet the definition of a forest: i.e., industrial plantations for timber, food, oil palm etc. Now it can be consider as a Forest Protected area.

In Wolong Nature Reserve, Sichuan, China a strict protected area, that falls in the IUCN category (Ia) i.e. strict nature reserve established primarily to protect the giant panda. At Sugarloaf Mountain, Brecon BeaconsNational Park, UK the woods on the side of the mountain are owned and managed as a protected area. In the same way countries declare forest protected based on the IUCN definition of a forest protected area (Dudley 2008).

In Pakistan, protected areas provide goods and services to a large number of people (Pakistan Forest Act 1927; Pakistan wildlife Act 1974, Jan 1992). Forest protected areas (initially declared for sustaining forest resources, but recently also oriented towards wildlife conservation) are divided into state-owned and private/community-owned categories: according to Jan (1992), 66% are state forest while 34% are owned by the local communities or privately (Worboys and Lockwood 2015).

(Bruno and Follador 2006) cited in "Integrating Remote Sensing, GIS and Prediction Models to Monitor the Deforestation and Erosion in Peten Reserve, Guatemala" about the strategy for modelling the deforestation by analyzing the land cover dynamics and by multi temporal analysis of Normalized Difference Vegetation and Hydric Stress index. The author points out the importance of new environmental criteria i.e. distance from roads, oil pipe-line, DEM, etc. These criteria effect the spatial allocation of predicted land cover probabilities and also compare the results of different approaches i.e. Markov Chains, Multi Criteria Evaluation and Cellular Automata; Neural Networks for improving the final method for the assessment of future deforestation risk.

(Bavaghar 2015) cited in "Deforestation modelling using logistic regression and GIS" that Geographic information system and logistic regression analyses were used to predict the deforestation. The factors contributing in forest degradation in Hyrcanian forests of western Gilan were also found. By using logistic regression model it has been found that deforestation is a function of slope, distance to roads and residential areas. In the deforestation assessment the role of slope factor is more than distance factor. This model determines that deforestation is negatively related to slope.

(Reveshty 2011), in this study Fuzzy art based supervised classification is used to identify land use modifications happened in the Zanjan, Iran. The CA and Markov models have been used for projecting land-use change till 2020. In this paper, different image processing techniques have been used on 1984, 1991, 2006 and 2011 Landsat TM images.

(Mas and Paegelow 2014) cited in "Modelling Tropical Deforestation: A Comparison of Approaches" Deforestation is an important factor of global warming. In last few decades this topic has received great attention by researchers and scientists. GIS based spatial models have been used worldwide for prediction land use cover change (LUCC). Many software's and approaches have been used to model (LUCC) such as

CLUE-S, DINAMICA GEOMOD and IDRISI. In this study the scientist have compared these four modelling approaches. In this study deforestation were modelled by using these four approaches and output maps were compared. For this comparison two land cover maps (date 1 and 2) were analyzed and four explanatory maps that are elevation, slope, distance to roads and distance to settlements were taken. For simulation three land cover classes were generated as forest, pasture land and shifting agriculture. In this study after comparison of these four approaches the author states that in GEOMOD the classes like deforestation o pasture and deforestation to agriculture were merged into single class i.e. deforestation. The author states that in IDRISI the suitability maps of each variable were elaborated using multi-criteria method. DINAMICA gives more realistic generated landscape.

(Arekhi 2011) cited in "Modeling spatial pattern of deforestation using GIS and logistic regression: A case study of northern Ilam forests, Ilam province, Iran" the factors contributing in forest degradation i.e. distance from road and settlement areas, forest fragmentation index, elevation, slope and distance from the forest edge of Northern forests of Ilam province. Forest cover change has been analyzed. Logistic regression model is used for estimating the spatial distribution of deforestation. The result of this modelling shows more deforestation in the fragmented forest cover and in the areas near to forest edge. The factors like distance to roads, slope and settlements has negative relationship with deforestation. Deforestation rate is increased with decrease in elevation.

Forest standards must be followed in order to save habitat and environment. Distance is an important factor in order to prevent deforestation and encroachments. In this study it has been found that which factors causes deforestation and how. There are many causes of deforestation. In this study, criteria or factors affecting suitability of a specific land-use type i.e. forest, includes bio-physical factors as slope, distance to road, distance to drainage network and distance to settlements. The physical proximity to an existing land-use class is thought to be a driver of change into a particular land-use class in future. These criteria or factors exists at recommended threshold values then the Forest is Fit else. It can be deforested or replaced by any other land-use type.

On the basis of mention criteria i.e. Slope, EVI, Forest Land use type and distance to roads, drainage network, and settlement, forest protection zones have been identified. Land use cover change models are important for understanding the global environmental change. Different prediction models have been used worldwide for land use prediction. Prediction analysis performed to analyze how the land use will change with the passage of time and how it affect the human life and earth. These prediction models help the decision makers that which land use changes brings good or adverse effects on environment. There are many approaches and software to model land use cover change (LUCC). In this study CA-Markov model has been used for spatial modelling of forest protection zones.

This study aimed to counsel a mechanism for identification of forest protection zones. The zoning is based on the forest fitness criteria and land use suitability. Procedure of ranking of different land use and other criteria has been established. The details of these steps will be explained in below sections. The next section will discuss about the study area and the material and methodology used in the study area. After this section result will be explained in context with the methodology. Finally the conclusion has been drawn based on these results.

2 Materials and Methods

2.1 Study Area

Murree is an administrative tehsil of district Rawalpindi of Province Punjab, Pakistan. It exists beyond the capital city, Islamabad. Geometrically Murree lies at 73°23′42.2″ E and 33°52′26.43″ N coordinates. Murree is an administrative center of Murree Tehsil. It is a mountainous region with an elevation of 2,291.2 m (7517.1 ft.) approximately. The study area has moderate to steep slopes. Coniferous and Scrub Forests exists in Murree. Coniferous Forests exists at higher elevation and Scrub is found throughout the Murree. In the east of Murree tehsil River Jhelum exists. It also includes the parts of Margalla hills around Islamabad. It is the famous hill station of Pakistan. Patriata and Ayubia are the pivotal attractions of Murree. Patriata is the highest point of Murree Hills and of the Punjab also. The study area has cold, snowy winters and relatively cool summers.

2.2 Methodology

The satellite images of Landsat were used to generate Land use and Land Cover maps of Murree. Landsat-5 TM, Landsat-7 ETM+ (path 150, row 36 and 37) with 30 m spatial resolution dated 20-09-1992, 16-09-2005 and 19-09-2009 have been downloaded from the official website of U.S. Geological Survey (USGS). Aster dem having 30 m spatial resolutions was used to drive slope map. Land scan data of 2015 was used to drive settlements in the study area. The boundary of Murree Tehsil was used.

Landsat TM images of year 1992, 2005 and 2009 were employed in this study to produce Land use and land cover categories respectively. The images were projected to WGS-1984 and UTM Zone-43 N Coordinate System. The images were classified using supervised classification method. Almost 25 to 35 training areas were selected for each class on the image, on the basis of these training areas signature files have been generated. The images were classified into three classes i.e. Forest, Barren Land and other Classes. The classes like built up, water body and roads were merged into single class i.e. named as Other Classes as shown in Figs. 1, 2 and 3. It is very important for the future Land use and cover map prediction that all the images needs to processed in a way to have all resolutions i.e. image, spectral, spatial and radiometric resolutions must be consistent.

In Markov model, the Land Use Land Cover (LULC) image of 1992 of Murree was used as the base image (t) whereas the LULC image of 2005 was used as the later image (t + n) to get the transition area matrix (TAM) between 1992 and 2005 for predicting the LULC image of 2009. In the same way the LULC image of 2005 was used as the base image (t) while the projected LULC image of 2009 was used as the later image (t + n) to generate the transition area matrix (TAM) and transition probability matrix (TPM) as shown in Table 3 and Table 4 between these years to predict

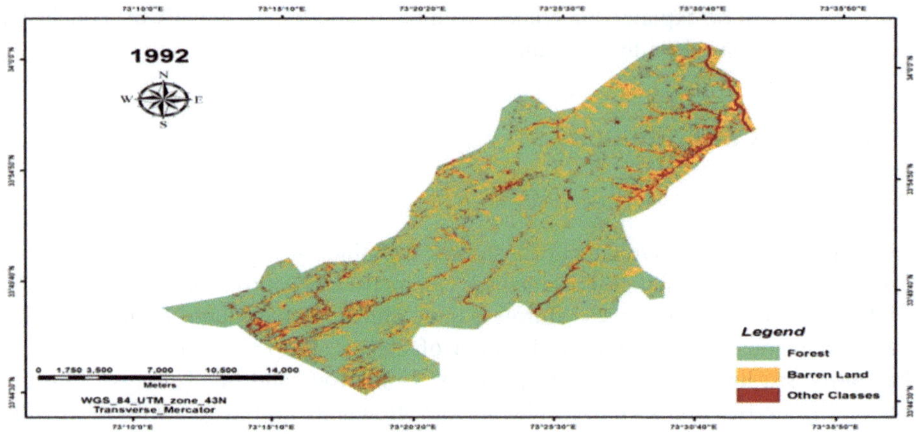

Fig. 1. Land Use Land Cover Map 1992

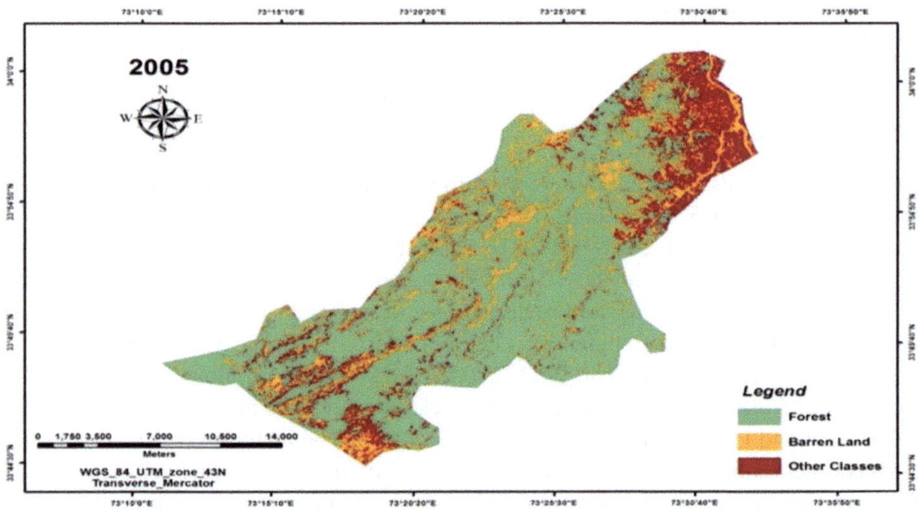

Fig. 2. Land Use Land Cover Map 2005

the future LULC image of 2020. These probabilities were generated from past changes for the future prediction. Although the Markov model is not suitable as it does not have spatial component. It may give the right magnitude of change but not the right direction (Ralph and Boerner 1996).

In Cellular Automate (CA) standard and user defined contiguity filters are used. The contiguity rule is based on this logic if a pixel near to agriculture land is most likely to change into agriculture land. In this study 5 × 5 mean contiguity filter was used to define the neighborhood relationship for projecting LULC of 2020. The transition area file and set of conditional probability images were used for projecting LULC

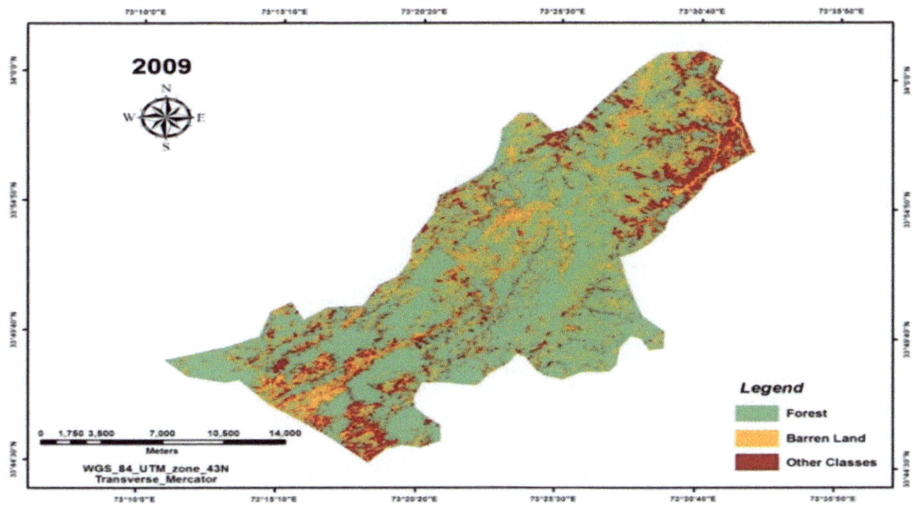

Fig. 3. Land Use Land Cover Map 2009

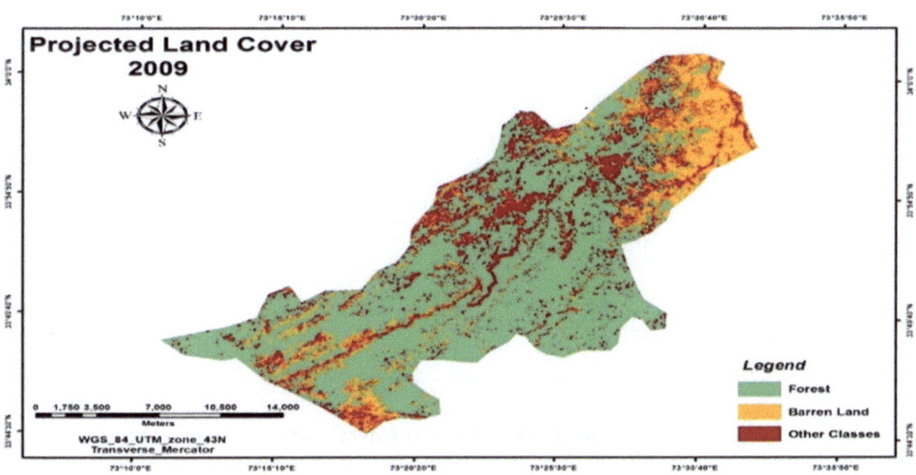

Fig. 4. Projected Land Use Land Cover 2009

of 2020. CA-Markov then begins an iterative process and runs11 number of iterations based on the number of time steps.

Spatial Validation among observed and projected image has been done for projecting the LULC of 2020.

Murree Forest Land Use Type Map was prepared from the base line report on Flora of Murree Biodiversity Park that has been published by Environment Protection Department (EPD), Punjab and IUCN Pakistan (Zafeer Saqib 2014). In this report the authors have mentioned the flora of Murree with respect to their Forest Types. The Forest Type map was prepared from the information about elevation given in this

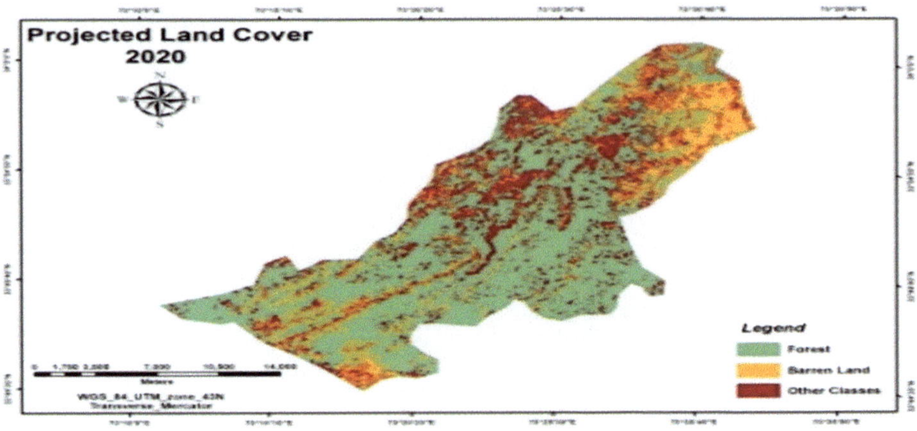

Fig. 5. Projected Land Use Land Cover 2020

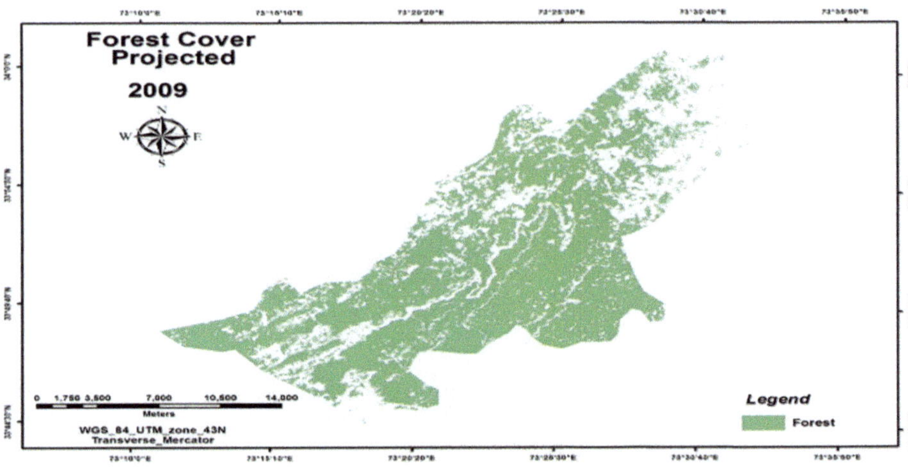

Fig. 6. Projected Forest Cover 2009

report. As per report the Sub-tropical Broad Leaved Forest type exists at 639–1050, Sub-tropical Chir Pine Forest exist at 1050–1800 and Moist Temperate Forest exist at 1800–2229 these elevation.

For the preparation of Drainage Network of study area, it was projected from ASTER-SRTM 30 m DEM.

For the preparation of roads map, the roads were digitized to analyze how this factor is causing deforestation in Murree tehsil.

The Enhanced Vegetation Index (EVI) map of Murree Tehsil was created using the following Eq. 1 and then EVI map has been classified into classes i.e. Low green areas (−0.33–0.40) and high green areas (0.40–0.72).

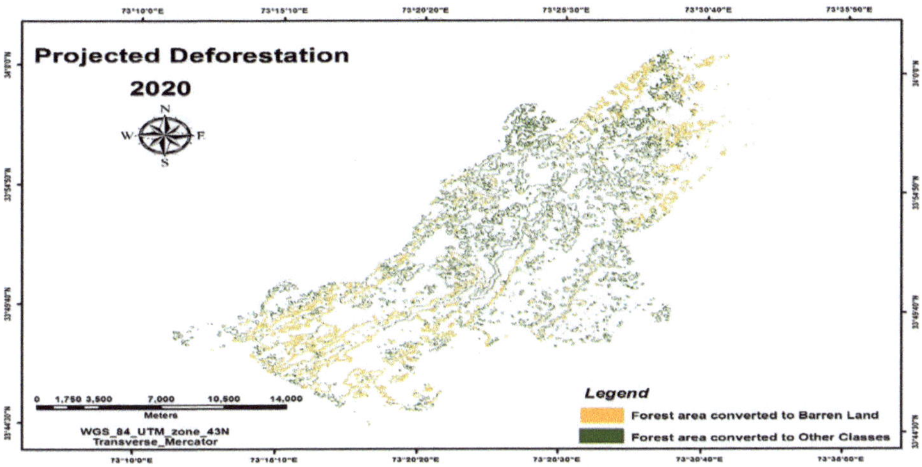

Fig. 7. Projected Deforestation in 2020

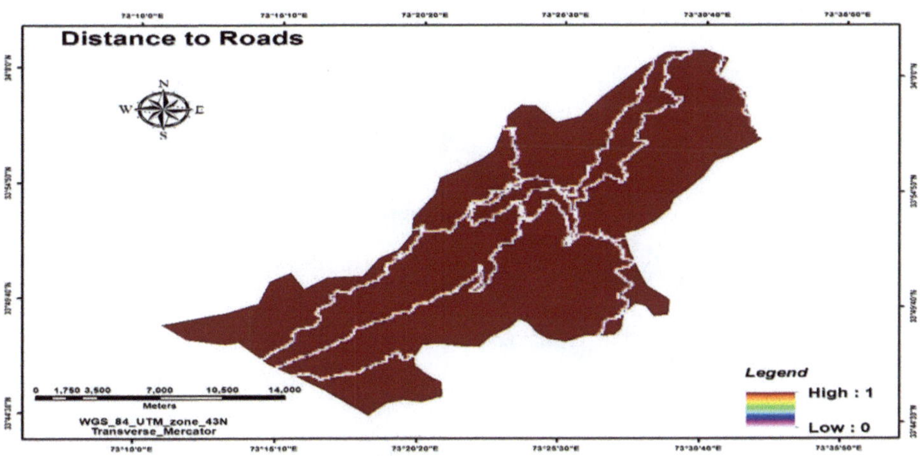

Fig. 8. Distance to Roads Map

$$EVI = G * \frac{(NIR - RED)}{NIR + C1 * RED - C2 * Blue + L)} \qquad (1)$$

Where

NIR/red/blue are surfaces reflectances
L is the canopy background adjustment that addresses non-linear
C1, C2 are the coefficients of the aerosol resistance term
G is the Gain Factor that is equal to 2.5

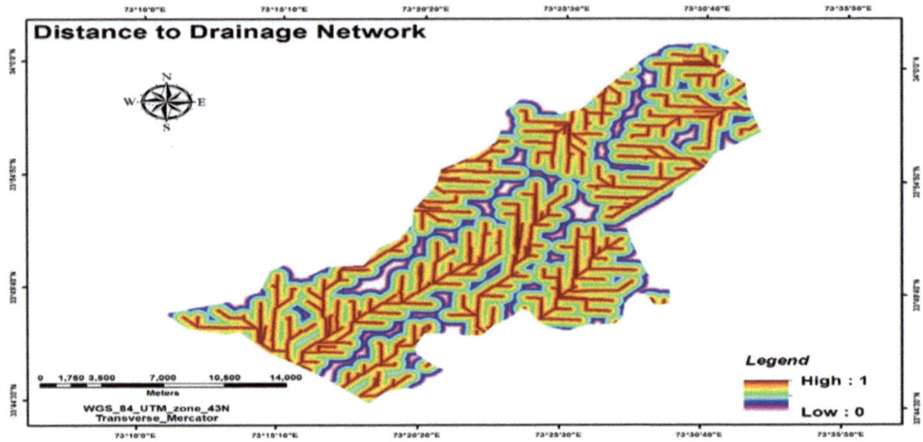

Fig. 9. Distance to Drainage Network Map

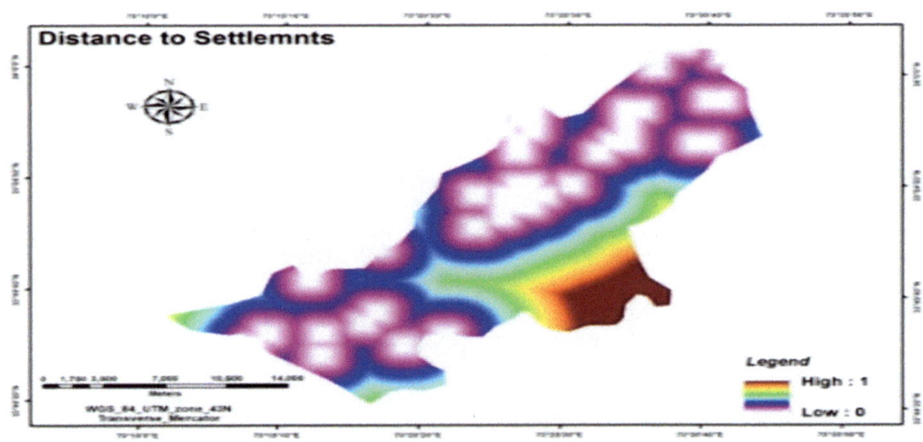

Fig. 10. Enhanced Vegetation Index Map

In this study Forest Fitness criteria were identified. The areas were pointed out that are most important to protect in future is based on the mentioned criteria. For each criterion a factor map was generated. Factor maps i.e. slope map, distance to drainage network, distance to roads, distance to settlements, Forest land use type map and EVI map. These factor maps will use as criteria for further analysis. Multi criteria analysis (MCA) is performed on these criteria maps. The result of MCA provides the importance (in percentage) of particular piece of forest at pixel level (30 m). A threshold value of 60 and 90% is applied on the resultant raster. This result in final is a forest protection zones map. The flow of the methodology has been shown in Fig. 19.

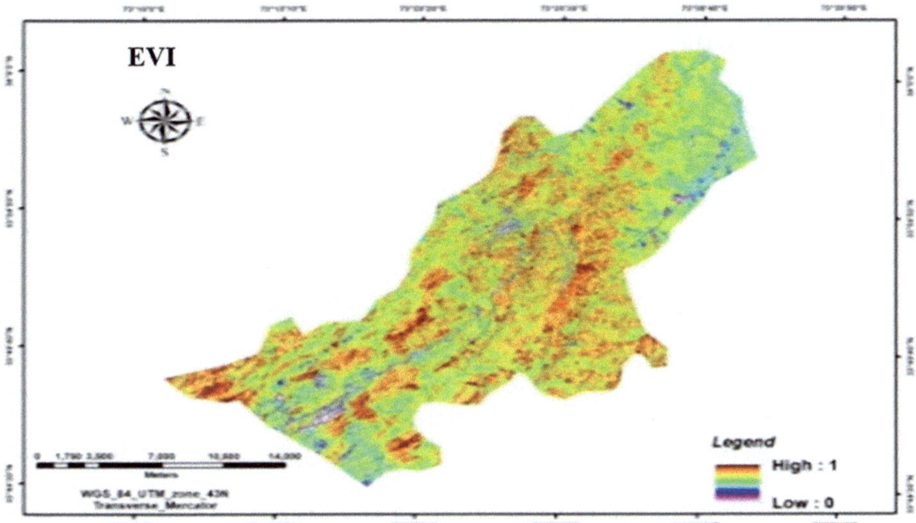

Fig. 11. Distance to Settlements Map

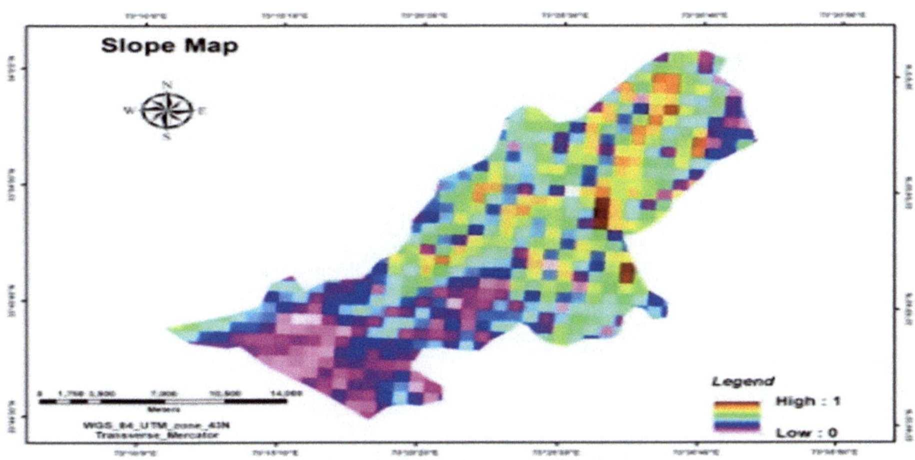

Fig. 12. Slope Map

3 Results and Disscussions

3.1 Land Use Land Cover Maps

The spatial extent of forest class in 1992, 2005 and 2009 was 74.6, 67.1 and 64.8% and the spatial extent of barren land in the above mentioned years were 19.8, 5.54 and 13.33 whereas in 1992, 2005 and 2009 there is a rapid increase in the spatial extent of

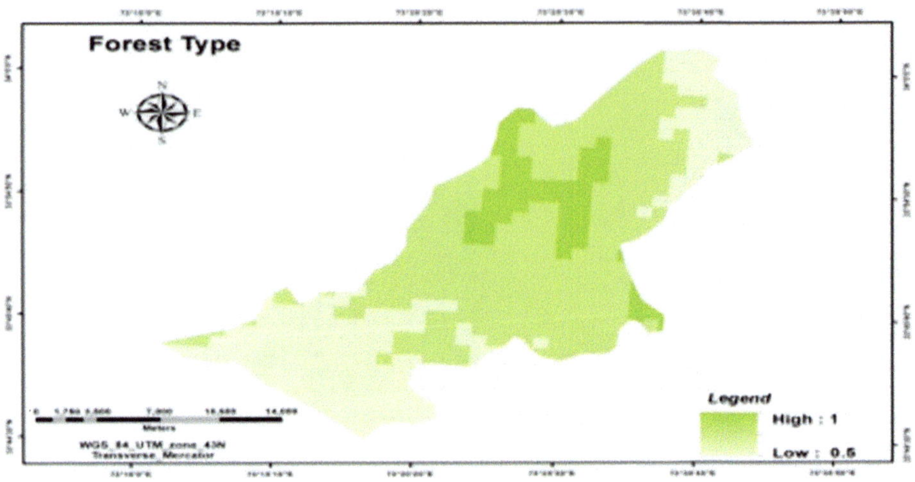

Fig. 13. Forest Land use Type Map

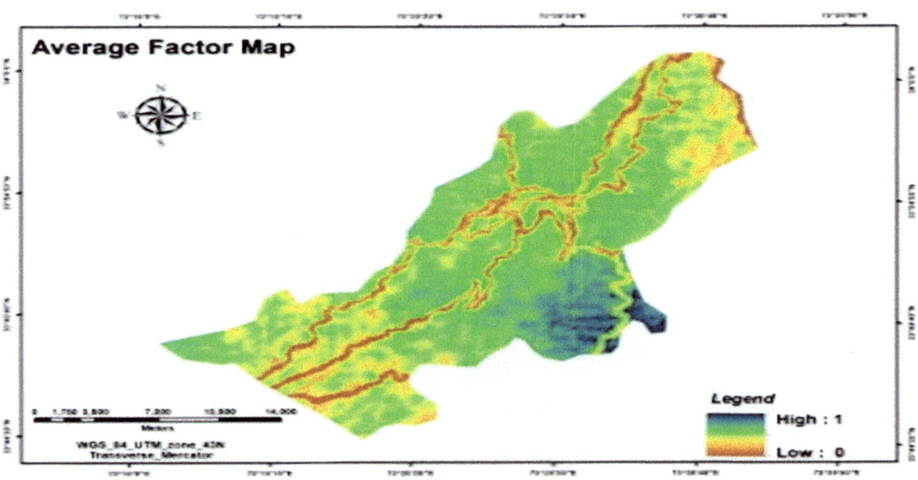

Fig. 14. Forest Protection Zones

other classes 5.4, 17.3 and 21.7%. The LULC images of respective year have been shown in Figs. 1, 2 and 3 and also discussed in Table 2.

3.2 CA-Markov

The basic framework of Markov is to observe probabilities of past trends organized into a matrix (Houet and Hubert-Moy 2006). This model produced Transition Probability Matrix (TPM), Transition Area Matrix (TAM) and a set of Conditional Probability images (CPM) for each LULC type by analyzing a pair of land cover images i.e.

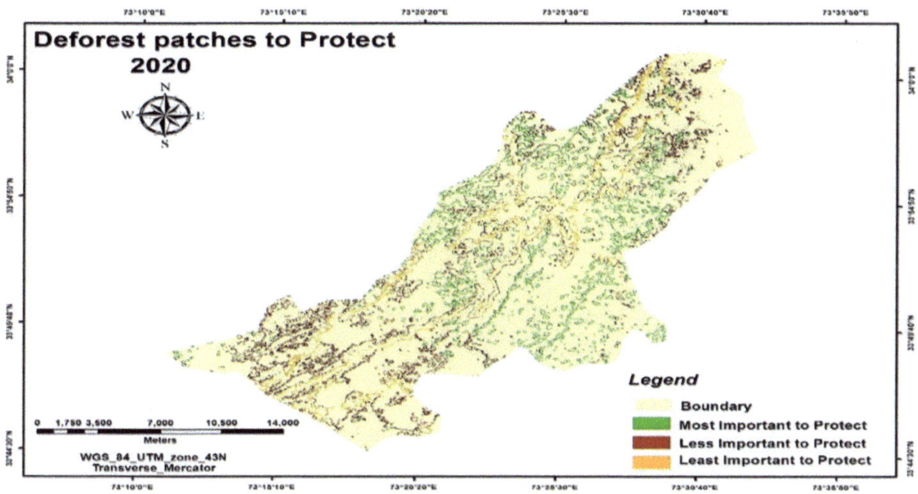

Fig. 15. Important deforest patches to Protect

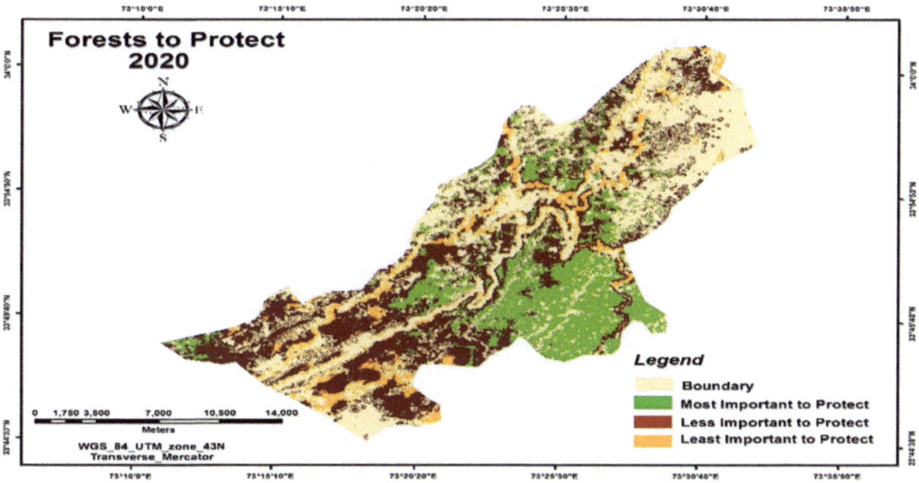

Fig. 16. Important Forests to protect

1992 and 2005 for predicting the LULC of 2009 and 2020 as shown in Tables 3 and 4. CA-Markov model interprets any number of land use categories based on a cell based matrix generated by Markov. CA-Markov model deals with both spatial and temporal changes (White and Engelen 2000). The projected LULC image of 2020 is shown in Fig. 5 and also discussed in Table 5.

Spatial Validation of Predicted 2009 image and Observed 2009 image was done by subtracting observed 2009 image from predicted 2009 image using the following Eq. 2.

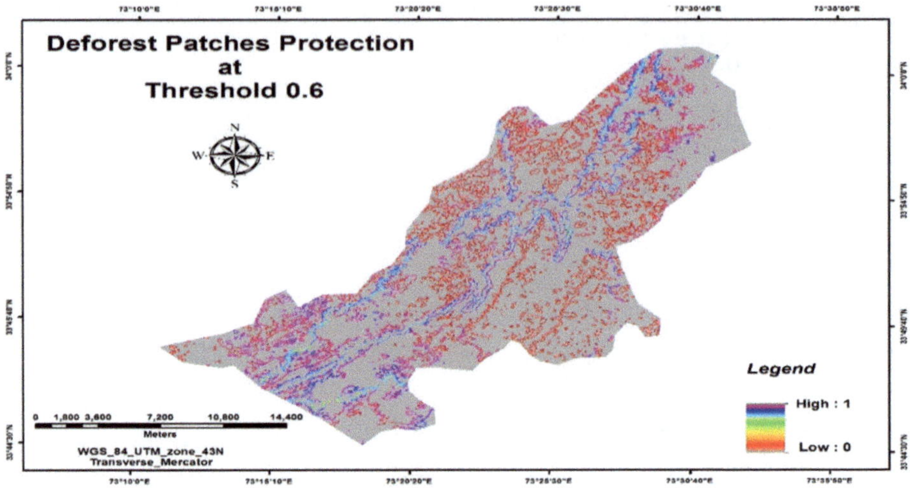

Fig. 17. Deforest Patches protection at 60%

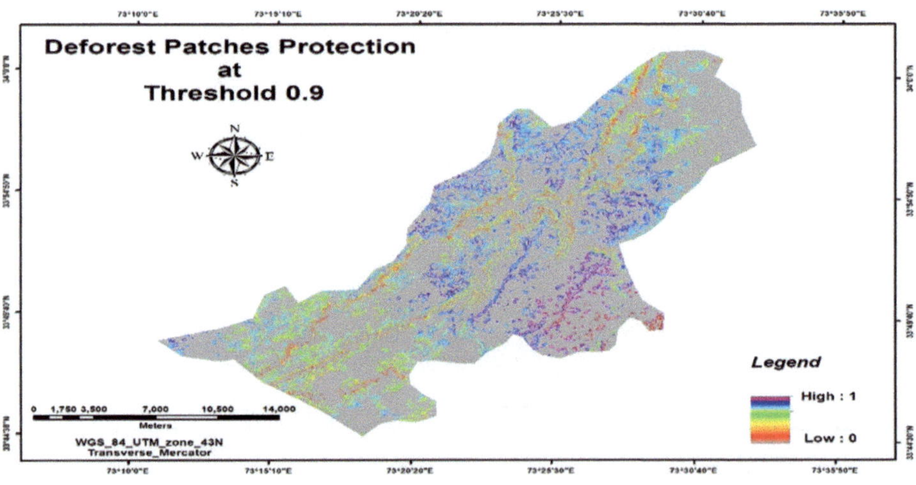

Fig. 18. Deforest Patches protection at 90%

$$\textbf{\textit{Spatial Validation}}_{\mathbf{2009}} = (\textbf{\textit{Predicted LULC}}_{\mathbf{2009}} - \textbf{\textit{Observed LULC}}_{\mathbf{2009}}) \quad (2)$$

Where as "***Observed LULC*$_{2009}$**" is a Land use land cover map of 2009 as shown in Fig. 3 and "***Predicted LULC*$_{2009}$**" is a projected LULC map as shown in Fig. 4.

The minimum target accuracy kept 60% for the prediction of 2020 LULC map but when the accuracy assessment was done on the both images it gives 68% accuracy. This accuracy was considered sufficient for moving towards the prediction of 2020 LULC map.

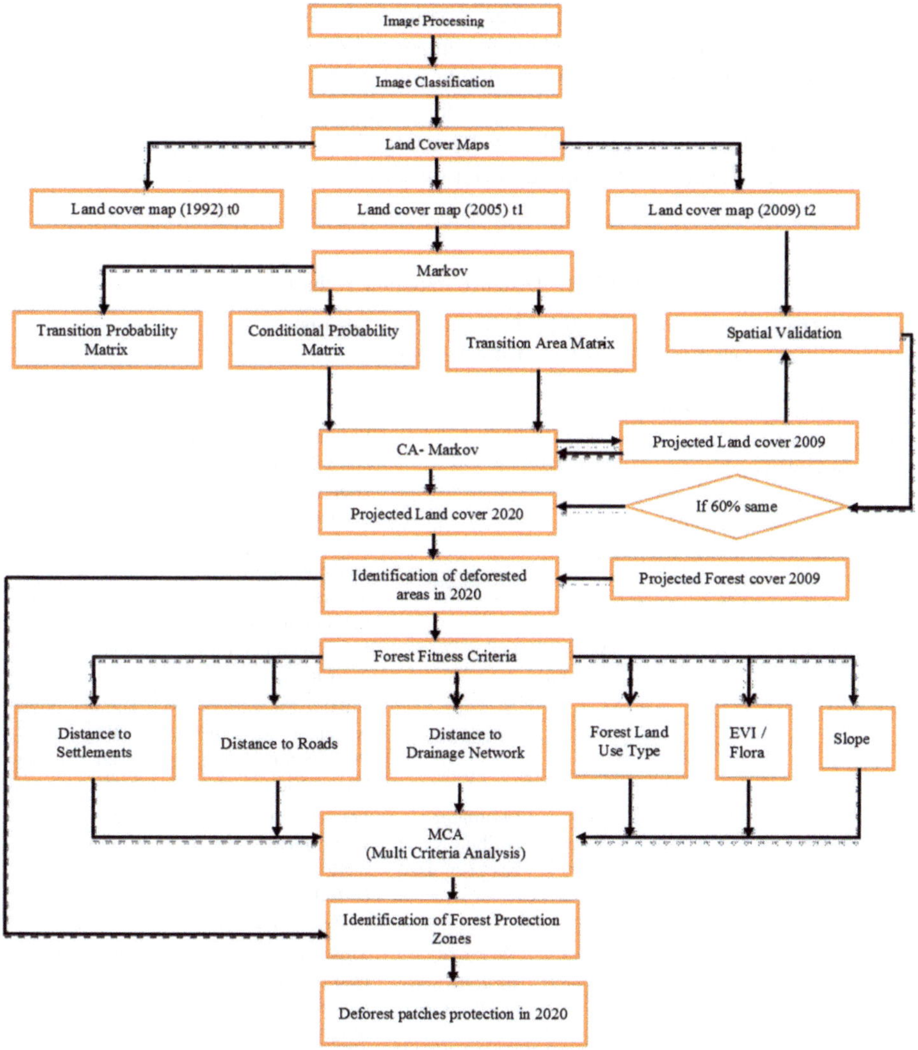

Fig. 19. Flowchart of Methodology

3.3 Forest Fitness Criteria

In this analysis, criteria or factor affecting suitability of a specific land-use type i.e. forest, includes bio-physical factors as slope, distance to road, distance to drainage network and distance to settlements. The physical proximity to an existing land-use class is thought to be a driver of change into a particular land-use class in future. As per Literature review of these criteria or factors exists at recommended threshold values then the Forest is said to be Fit else it can be deforested or replaced by any other land-use type.

A distance importance chart has been developed by literature review as shown in Table 1 displaying the threshold values for creating factor maps. Most Suitable means the weightage is maximum and least suitable means the assigned weightage is minimum. As per Literature review there are many factors causing deforestation. These criteria or factors maps were developed by using different datasets and methods as discussed earlier. Each factor map has been discussed below in detail.

Table 1. Distance importance chart

Theme	Criteria	Maximum attribute values (meters)	Verbal scale	Recommended threshold value	Description
Forest	Distance to road	6185.1	Most suitable	70 m	For the junction between the Public roads and forest roads the standards must be adopted. For the national routes the sight distance can be 90 m and for the regional routes it must be 70 m. As every forest area does not exit on a public road so for the sake of minimizing hazards special measurements must be taken (Tom Ryan)
	Slope	19.386°	Most suitable	18°	For soil suitability the slope above 18° must be forested. The convex slope is suitable for forested land. The key point is suitable for settlements. (DIY Permaculture Landscape Design: Making Use of Slope and Aspect 2015
	Distance to drainage network	1755.7	Least Suitable	Minimum attribute value has been used	Forests reduce soil erosion and maintain water quality. The major cause of increase in soil erosion is deforestation. (I. Calder, Towards a new understanding of forests and water)
	Distance to settlements	11015.6	Most suitable	Maximum attribute value has been used	This study has proved that development near protected areas decrease the value of conservation and also not good for biodiversity. (Hosseini and Khalili 2012)
	Flora/EVI	0.72477	Most suitable	Maximum attribute value has been used	The areas with high green values needs to be protected.

Note: Maximum Attribute Value means the value of attribute table of spatial layers

Table 2. Land use land cover classification

Categories	1992 (hc)	%	2005(hc)	%	2009(hc)	%
Forest	28556.55	74.65	25672.86	67.11	24822.81	64.89
Barren Land	7599.42	19.86	5946.75	15.54	5100.84	13.33
Other Classes	2100.24	5.49	6636.6	17.35	8332.56	21.78
Total	38256.21		38256.21		38256.21	

Table 3. TAM (2005–2009)

Given: Probability of changing to:			
	Forest	Barren land	Other Classes
Forest	208124	27662	39075
Barren land	28676	19025	20045
Other classes	19534	29300	33629

Table 4. TPM (2005–2009)

Given: Probability of changing to:			
	Forest	Barren land	Other classes
Forest	0.7572	0.1006	0.1422
Barren land	0.4233	0.2808	0.2959
Other classes	0.2369	0.3553	0.4078

3.3.1 Distance to Roads Map

Distance map was generated from each pixel of the study area to access that which deforest patches are near to roads and which deforest patches are far away from this Land-use type.

The normalized map of this factor is shown in Fig. 8 prepared by using threshold value 70 m. By using the following Eq. 3 the deforested patches that are more than 70 m distances were assigned the weightage of 100% because they are important to protect.

$$f(Edis_r) = \begin{cases} \frac{Edis_r}{70}, & Edis_r < 70 \\ 1, & Edis_r \geq 70 \end{cases} \tag{3}$$

Where as "**Edis$_r$**" is a distance to roads map as shown in Fig. 8 and 70 m is a threshold value as discussed in Table 1.

3.3.2 Distance to Drainage Network Map

Drainage Network Distance map is shown in Fig. 9 was generated from each pixel of the study area to access that which deforest patches are near to waterbodies and which deforest patches are far away from this Land-use type. The weighted map of this factor map was developed by using the Eq. 4. The areas near to drainage network are more important as compare to those deforest patches that are far away from drainage network.

$$f(Edis_{dn}) = \begin{cases} \frac{Edis_{dn}}{1755.7}, & Edis_{dn} \geq 1755.7 \\ 1, & Edis_{dn} < 1755.7 \end{cases} \tag{4}$$

Where as "**Edis$_{dn}$**" is a distance to drainage network map as shown in Fig. 9 and 1755.7 m is a maximum attribute value as discussed in Table 1.

3.3.3 Distance to Settlements Map

Distance to settlements from each pixel of study area was calculated. This Factor map was developed and normalized by using the Eq. 5 as shown in Fig. 11. The areas near to settlements are less important as compare to the areas that are far away from settlements. 100% weightage is assigned to those patches/areas that are important to protect and vice versa.

$$f(Edis_{settlements}) = \begin{cases} \frac{Edis_{settlements}}{11015.6}, & Edis_{settlements} < 11015.6 \\ 1, & Edis_{settlements} \geq 11015.6 \end{cases} \quad (5)$$

Where as "$Edis_{settlements}$" is a distance to settlements map as shown in Fig. 11 and 11015.6 m is a maximum attribute value as discussed in Table 1.

3.3.4 EVI/Flora Map

The EVI map was normalized using the following Eq. 6 in raster calculator as shown in Fig. 10.

$$f(EVI_{2009}) = \begin{cases} \frac{EVI_{2009}}{0.72477}, & EVI_{2009} < 0.72477 \\ 1, & EVI_{2009} \geq 0.72477 \end{cases} \quad (6)$$

Where "$\mathbf{EVI_{2009}}$" is an EVI map as shown in Fig. 10 and 0.72477 is a maximum attribute value as discussed in Table 1.

3.3.5 Slope Map

The prepared slope map was normalized using the following Eq. 7 as shown in Fig. 12. The 100% weightage was assigned to areas that have slope greater than 18° and vice versa.

$$f(Edis_{slope}) = \begin{cases} \frac{Edis_{slope}}{19}, & Edis_{slope} < 19 \\ 1, & Edis_{slope} \geq 19 \end{cases} \quad (7)$$

Where as "$\mathbf{Edis_{slope}}$" is a slope map as shown in Fig. 12 and 19° is a maximum attribute value as discussed in Table 1.

3.3.6 Forest Land Use Type Map

The Forest type map was normalized by assigning the 0.5 pixel values to the Sub tropical Broad Leaf Forest Type, 0.75 pixel values was assigned to the Sub tropical Chir-pine Forest Type and 1.0 was assigned to the Moist Temperate Forest as shown in Fig. 13.

3.4 Forest Protection Zones

Forest Protection Zones have been identified by using MCA. The above mentioned factors map i.e. Distance to Roads, Distance to settlements, Distance to drainage

Table 5. Projected land cover 2020

Category	Area (hc)	%	Category	Area (hc)	%	Category	Area (hc)	%
Forest to Forest	208124	81.2	Forest to Barren Land	27662	36.4	Forest to Other Classes	39075	42.1
Barren Land to Forest	28676	11.2	Barren Land to Barren Land	19025	25.0	Barren Land to Other Classes	20045	21.6
Other Classes to Forest	19534	7.6	Other Classes to Barren Land	29300	38.6	Other Classes to Other Classes	33628	36.3
Total Forest (cells)	256334	-	Total Barren Land (cells)	75987		Total Other classes (cells)	92748	-
Total Forest (hc)	23070.06	60.3	Total Barren Land (hc)	6838.83	17.9	Total Other classes (hc)	8347.32	21.8
Total Murree Tehsil (hc)					38256.21			

Table 6. Projected deforestation 2020

OID	Value	Detail	Count	Area (hc)	%
0	1	Forest Area converted to Barren Land	27662	2489.58	6.59
1	2	Forest Area converted to Other Classes	39075	3516.75	9.19
Total area:-			425069	38256.21	

Network, EVI map and Forest land use type map were added and then average of these six factors were taken using raster calculator by using the following Eq. 8.

$$Forest\ Protection\ zones = \frac{EDIS_r + EDIS_{settlements} + EDIS_{dn} + EVI + Forst\ Type + EDIS_{slope}}{6} \tag{8}$$

Where

$EDIS_r$ = distance to roads map as shown in Fig. 8
$EDIS_{settlements}$ = distance to settlements map as shown in Fig. 11
$EDIS_{dn}$ = distance to drainage network map as shown in Fig. 9
EVI = Enhanced Vegetation Index map as shown in Fig. 10
Forst Type = Forest Type Map as shown in Fig. 13
$EDIS_{slope}$ = distance to slope map as shown in Fig. 12

The output map as shown in Fig. 14 is the suitability map for Forest. This map is showing the most suitable areas for forest land use class. These areas need to be highly protected so they can be protected from deforestation in future. The Fig. 14 is showing the areas that are highly feasible for the forest conservation as per the six criteria.

3.5 Identification of Deforested Areas

Deforested areas were identified by using the following Eq. 9. The output map of deforested areas is shown in Fig. 7 along with the detail of forest class conversion to Barren land and other classes as shown in Table 6.

$$\textbf{\textit{Predicted Deforested Patches}}_{\mathbf{2020}} = (\textbf{\textit{Projected LULC}}_{\mathbf{2020}} - \textbf{\textit{Projected FC}}_{\mathbf{2009}}) \quad (9)$$

Where the "***Projected LULC*** $_{\mathbf{2020}}$" is a map shown in Fig. 5 and "***Projected FC*** $_{\mathbf{2009}}$" is a map of projected Forest cover of 2009 as shown in Fig. 6.

3.6 Important Deforested Patches Protection

The output map of important deforest patches has been generated using given Eq. 10 as shown in Fig. 15. Both maps i.e. predicted deforested patches 2020 map and Forest Protection Zones map have been multiplied with each other. So that it can be viewed or observed that at which places deforested patches are important to protect.

$$\textit{Deforest patches protect}_{2020} = (\textit{Predicted deforest patches}_{2020} * \textit{Forest Protection Zones}) \quad (10)$$

Where the "***Predicted deforest patches*** $_{\mathbf{2020}}$" is a map as shown in Fig. 15 and "***Forest Protection Zones***" is a map as shown in Fig. 14.

3.7 Protection Level at Specific Threshold Values

The Forest Protection levels have been accessed by using MCA. As there is no yardstick to measure that at what specific threshold value the deforest patches are important to protect. So, the predicted deforest patches and other areas have been reclassified. This reclassified map i.e. predicted deforest patches was multiplied by protection zones map and then divided by different threshold values i.e. 60 and 90%.

$$\textit{Deforest patches protect}_{60} = \frac{(\textit{Predicted deforest patches}_{2020} * \textit{Forest Protection Zones})}{60} \quad (11)$$

$$\textit{Deforest patches protect}_{90} = \frac{(\textit{Predicted deforest patches}_{2020} * \textit{Forest Protection Zones})}{90} \quad (12)$$

Where the "***Predicted deforestpatches*** $_{\mathbf{2020}}$" is a map as shown in Fig. 15 and "***Forest Protection Zones***" is a map as shown in Fig. 14.

The deforest patches protection at threshold 60% is shown in Fig. 17 and the protection at threshold 90% is shown in Fig. 18.

3.8 Important Forests to Protect

The output map of important forests was generated using given Eq. 13 as shown in Fig. 16, projected forest 2020 map and forest protection zones map have been multiplied with each other. So that it can be viewed or observed that which forests are important to protect.

$$Forests\ to\ protect = (Projected\ Forest\ Cover_{2020} * Forest\ Protection\ Zones) \quad (13)$$

Where the "***Projected Forest Cover*** $_{\mathbf{2020}}$" is a map of projected Forest Cover of 2020 and "***Forest Protection Zones***" is a map as shown in Fig. 14.

4 Conclusions

Thematic land use and land cover types of 1992, 2005 and 2009 of Landsat sensor have been classified. The spatial extent of forest cover in 1999, 2005 and 2009 has been found 74.65%, 67.11% and 64.89%. It has been cleared from the LULC maps of 1999, 2005 and 2009 that the forest cover has been decreasing.

In order to predict areas with high risk for future deforestation; the analysis has been performed by a remote sensing based land cover prediction model(s) i.e. Markov and CA-Markov. By using these models LULC map of 2020 has been projected and it has been observed that forest cover will be left 60.3% by 2020. The predicted LULC map of 2020 will help decision makers in planning.

This study aimed to counsel a mechanism for forest conservation. In this study Forest Fitness criteria have been found. Each criteria has its own importance. The forests located at lower elevation and close to roads have been observed most vulnerable to clearance. Distance to roads and settlements are explored as the main driving force for loss of forest.

Distance importance chart has been developed to check that at how much distance any land use type should exist from Forest Land use class by generating distance map of each factor. Which land use type is beneficial for Forests to exist near or vice versa? This chart is based on logics and threshold values that have been taken from literature review.

On the basis of Forest Fitness Criteria, the Forest Protection Zones have been pointed, the pockets that are most susceptible to deforestation in 2020 have been identified and forests have also been mapped that are most important to protect in future.

Deforest patches have been identified in the study area. Important deforest patches protection has been assessed at different threshold values i.e. 60 and 90%. At 90% threshold value the deforest patches are most important to protect and these patches exists at same location of Forest Protection Zones also. The patches at 90% threshold value are the most important to protect as compare to 60% threshold value.

Geographical Information System (GIS) with the combination of Markov and CA models implemented effectively. In this study six criteria have been used for the identification of Forest protection zones, other criteria like Land sliding and Fauna can also be used to achieve desire results. Different authors have suggested different threshold values for each factor. The use of right threshold value is necessary for achieving desired outputs. The main challenge of the modelling is the normalization of factor maps using the Multi criteria analysis (MCA).

References

Anonymous: FRA 2000 on definitions of forest and forest change. Working Paper 33. Fao Corporative Document Repositiry, Rome (2000)

Anonymous: DIY Permaculture Landscape Design: Making Use of Slope and Aspect. Retrieved from DIY Permaculture Landscape Design: Making Use of Slope and Aspect, 02 November 2015. http://www.smallfarmpermacultureandsustainableliving

Arekhi, S.: Modeling spatial pattern of deforestation using GIS and logistic regression: a case study of northern Ilam forests, Ilam province, Iran. Afr. J. Biotechnol. **10**(72), 16236–16249 (2011). doi:10.5897/AJB11.1122

Bavaghar, M.P.: Deforestation modelling using logistic regression and GIS. J. For. Sci. **61**, 193–199 (2015). doi:10.17221/78/2014-JFS

Zafeer Saqib, I.K.: Murree Biodiversity Park-Baseline report on Flora. Environment Protection Department (EPD), Punjab and IUCN Pakistan, Islamabad, Pakistan (2014)

Dudley, N.: Guidelines for Protected Area Management Categories. IUCN, Gland, Switzerland (2008). https://iucn.oscar.ncsu.edu/mediawiki/images/1/1f/Dudley%282013%29.pdf

Worboys, G.L., Lockwood, M.: Protected Areas Governanace and Management. ANU Press (2015). https://books.google.com.pk/books?id=06WUCgAAQBAJ&pg=PA793&dq=forest+protected+area&hl=en&sa=X&ved=0ahUKEwjVwrfz9rfRAhUHWRoKHZ1QD70Q6AEIPjAH#v=onepage&q=forest%20protected%20area&f=false

Mas, J.F., Paegelow, M.: Modelling Tropical Deforestation: A Comparison of Approaches, 3 p., 12 September 2014. https://hal.archives-ouvertes.fr/halshs-01063568/document

Dudley, N., Parish, J.: Closing the Gap Creating ecologically representative. The Secretariat of the Convention on Biological Diversity (2006). http://areasprotegidas.info/upload/document/cbd-ts-24.pdf

Bruno, R., Follador, M.: Integrating Remote Sensing, GIS and Prediction Models to Monitor the Deforestation and Erosion in Peten Reserve, Guatemala. Society for Mathematical Geology, XIth International Congress, p. 6, 3rd to 8th September 2006. https://arxiv.org/ftp/arxiv/papers/0904/0904.0317.pdf

White, R., Engelen, G.: High-resolution integrated modelling of the spatial dynamics of urban and regional systems. Comput. Environ. Urban Syst. **24**, 383–400 (2000)

Ralph, E.J., Boerner, M.N.: Markov models of inertia and dynamism on two contiguous Ohio landscapes. Geograph. Anal. **28**, 56–66 (1996)

Reveshty, M.A.: The assessment and predicting of land use changes to urban area using multi-temporal satellite imagery and GIS: a case study on Zanjan, Iran (1984–2011). J. Geograph. Inf. Syst. **3**, 298–305 (2011). doi:10.4236/jgis.2011.34026

Hosseini, S.A., Khalili, M.: Relationship between forest road length and village development during three decades in Neka-Zalemrood forest. J. For. Sci. **58**, 1–7 (2012)

Houet, T., Hubert-Moy, L.: Modeling and projecting land-use and land-cover changes with Cellular Automaton in considering landscape trajectories: an improvement for simulation of plausible future states. EARSeL eProc. **5**, 63–76 (2006)

Wilkie, M.L.: Global Forest Resources Assessment 2010 Terms and Definitions. Forestry Department. Food and Agriculture Organization of the United Nations, Rome (2010). http://www.fao.org/docrep/014/am665e/am665e00.pdf

Road Traffic Accidents Viz-a-Viz Terrorism, Pakistan's Perspective

Muhammad Jawed Iqbal[1(✉)], Nasir Ali[1], Zaeem Sheikh[1], and Sadaf Saleem[2]

[1] National Institute of Transportation, National University of Sciences and Technology, Islamabad, Pakistan
[2] Freelance Medical Researcher, Islamabad, Pakistan

Abstract. Today road traffic accidents are a global health concern. Injuries due to road traffic accidents are the eighth leading cause of death globally, and the leading cause of death for young people aged 15–29. Approximately, 1.24 million Road Traffic Fatalities (RTF) occur annually and another 20 to 50 million sustain nonfatal injuries as a result of Road Traffic Crashes (RTC). These injuries and deaths have an immeasurable impact on the families affected, whose lives are often changed irrevocably by these tragedies, and on the communities in which these people lived and worked. The cost of dealing with the consequences of these road traffic crashes runs to billions of dollars.

On the other hand terrorism is considered as a major threat to society. During recent years, terrorism in Pakistan has become a major and highly destructive phenomenon. Terrorism is considered as a momentary calamity with a disastrous impact over a long period of time which claims lives, leaves people disabled, homeless and bankrupt. It is a disorderly occurrence, the pattern of which is unobservable.

This study investigates the fatality rate caused by Terrorist Attacks and Road Traffic Accidents (RTAs) in Pakistan. The data includes death rate and the number of occurrence of each incident for the time period of 12 years, i.e. from 2004 to 2015. Deaths caused by RTAs were revealed to have uniformity along the timeline whereas terrorism was shown to have very scattered data. The study concludes that although government considers terrorism as a critical issue and is trying its best to control it, it is difficult to eliminate terrorism due to complexity of the factors involved. On the other hand, the government is not serious in reducing road traffic injuries which is otherwise comparatively easy and the damages associated in terms of death toll and financial loss is too much. The paper is concluded with recommendations to curb RTAs and RTFs.

Keywords: Road traffic accidents (RTA) · Terrorism · Pakistan

1 Introduction

Terrorism is the use of violence or threat of violence in order to ingress a religious political or ideological alteration, mostly committed by non-state actors or undercover personnel or non-state actors serving on behalf of their particular governments. It is both mala in se (i.e., crime that is immoral or wrong in themselves) and mala prohibita

J. Calautit et al. (eds.), *Towards Sustainable Cities in Asia and the Middle East*,
Sustainable Civil Infrastructures, DOI 10.1007/978-3-319-61645-2_15

(i.e., crime that is made illegal by legislation). Religiously motivated terrorism is considered the most alarming terrorist threat today. The frightening terrorism stream in Pakistan can be attributed to a lot of reasons. These include explosive population growth, poverty, unemployment, Islamic religious schools, and a lack of vocational training Institutes in the country. The population of Pakistan is on a rising trend. An annual increase of 2% in the population of Pakistan has been estimated in the last 3 years [1]. In the midst of Islamic religious arguments, people do not allow for the use of family planning techniques. Many also believe that children are the poor man's only assets. Further, several families wish to have sons disregarding daughters for being burdensome.

An accident can be defined as "an unforeseen and unfavorable event, a bad luck unanticipated and with no evident cause." Most of the accidents are clashes that might and must have been evaded. There are a lot of varied classes of traffic accidents that import in severe personal injuries and fatalities.

When one vehicle front strikes a fixed object or another vehicle front it is called a head-on accident. This occurs when from opposite side of the avenue one vehicle enters a lane. This can also result when a car comes from the wrong route into a driveway, one-way street or expressway. A rollover accident occurs when a vehicle turns over on top of either the side or the top of the vehicle. This is due to harsh turning while at high speed. A smaller car has less chance to flip over than giant vehicles such as trucks or buses. Another accident in which only one vehicle is involved and the vehicle runs off of the street is called runoff accident, and normally occurs when driver turns the vehicle with the intention to avoid striking animal or another car in the path or when the driver is not giving attention. However, there are a lot of causes associated with these accidents.

Accidents can also occur due to equipment break down such as tire blowouts, brake fails, and steering breakdown. That's why car manufacturers are required to engineer and plan cars that provide driver safety in terms of car operations as well.

Some of the motor vehicle casualties may be due to roadway preservation, but not to the level that some drivers use it as a defense. Unluckily measures and safeguarding programs differ to a great extent from area to area and town to town, so nationally principles don't subsist. A cause of major accidents is over speeding as the hazard of a mishap is higher when the speed of a vehicle is higher.

Road Traffic Accident (RTA) reductions are the major and primary aspect of Transportation safety. Road Traffic Accidents (RTAs) are defined as "collision of a vehicle moving along a roadway with another vehicle or object. It may also include topping over of a vehicle". A large amount of injuries of both psychological as well as physical nature arise as a result of RTAs. About 1.25 million people die each year as a result of RTAs [2]. 90% of the world's fatalities on the roads occur in low and middle-income countries, even though these countries have approximately half of the world's vehicles [2].

Terrorism has also risen as a leading cause of fatalities. A total of 190,198 deaths occurred during 2006–2015 out of which 28,328 were in 2015 alone [3]. A far menacing and deadly cause of fatalities, terrorism has taken the focus of remedial and safety measures away from RTAs.

This study investigates the fatality rate caused by Terrorist Attacks and RTAs in Pakistan. The data includes death rate and the number of occurrence of each incident for the time period of 12 years i.e. from 2004–2015. The data of RTAs is acquired from World Health Organization (WHO) whereas terrorism fatalities data is acquired from South Asian Terrorism Portal and Pakistan Bureau of Statistics. The trends of accident and terrorism fatalities will be analyzed and subsequent recommendations will be provided.

2 Literature Review

The past literature focuses mainly on RTAs of Pakistan. Little to no literature is available pertaining to any terrorism fatalities studies carried out in Pakistan. National Injury Survey of Pakistan (NISP) conducted its first study in 1997 with its publications concentrated on traffic related injuries. The survey results showed that the overall injury rate was 41 per 1000 in a year out of which the share of traffic related injuries was 15 persons per 1000. The ratio to traffic injury exposure was found higher in males, particularly labors with the age ranging from 16–45 [4].

Bishail et al. (2003) conducted a study in Pakistan in which they concluded that the reduction in morbidity and mortality will be higher with high investment in road safety [5].

The present global burden of road traffic accidents is excessively incurred by low-income and middle-income countries. These countries are assumed to least afford the basic societal services. Majority of the road traffic victims in such countries comprise of vulnerable users in the form of pedestrians and motorcyclists, on which the interventions fail to focus. A recent study in Karachi showed that about 43% of people affected from RTAs are motorcyclists. *Stephen Luby, Imran Hassan* and others carried out a study on disproportionate role of striking vehicles and resulting injuries. They determined that 727 road injuries were reported in two emergency centers of Karachi over the period of three months from December 1993 to February 1994. Their results showed that out of those 727 persons, the share of non-fatality injured victims was 54% and deceased was 47%. The age of the deceased ranged from 20 to 40 years with the mean of 31 years it was reported that 8.7% of the victims were already dead when arrived the hospital, 9.4% died instantly during the treatment and 18% of them died later during the study period. It was analyzed that motorcyclists and pedestrians comprised of 13% and 35% of the total victims. The striking vehicles were commercial in 65% of total deaths and 49% of total non-fatal injuries. The relative risk of death by being hit by a commercial vehicle is 1.7 times higher than by other vehicles [6].

In 2011, a research was published under the authorship of *JA Razzak.* They discovered that a total of 99,272 victims of Road Traffic Injuries were registered in 5 big emergency departments of Karachi over the time period of September, 2006 to September, 2009. The mortality rate caused by accidents was calculated to be 5.7 per 100,000 persons whereas injury rate was 184.3 per 100,000 [7]. Another study in PIMS Islamabad showed that the RTAs had caused 281 deaths. About 63% of deceased were below the age of 40 [8].

The victims of traffic fatalities can be classified on the basis of epidemiological WHO sub-regions or on the basis of economic regions. Fatalities in African sub-region are 55% whereas is 15% or less in American or European regions. In low income countries pedestrian road traffic fatalities are 45% (227835 estimated), 29% (161,501 estimated) in middle income and 18% (22,500 estimated) in high income countries [9].

A report on international terrorism by Jewis Virtual Library stated that a total of 11,774 incidents had occurred worldwide in the year 2015 causing more than 28,328 fatalities and 35,320 injuries. The report also stated that between 2014 and 2015 the terrorist attacks are reduced by 13% and deaths caused by terrorist attacks are reduced by 14% [10].

According to a paper "The Global terrorism Database: Accomplishments and challenges" by Gary Lafree, the largest number of fatalities caused by terrorism in years 1970–2007 was 18000 people in Iraq [11].

3 Data Acquisition

The data for RTAs was acquired using Pakistan Bureau of Statistics [12]. Aggregate as well as disaggregate data were available on the website. The data for total injuries, fatalities as well as accident occurrence is used as is shown in Table 1 while data of individual province is shown in Table 2.

Table 1. RTA data of Pakistan

Year	Accidents	Killed	Injured
2004	9896	5112	12401
2005	9492	4868	11415
2006	10466	5465	12875
2007	10466	5615	12096
2008	9496	4907	11037
2009	9747	5280	11173
2010	9723	5271	11383
2011	9140	4758	10145
2012	8988	4719	9710
2013	8885	4672	9864

Table 2. Fatalities in provinces of Pakistan

Year	Sindh	Punjab	Khyber-Pakhtunkhwa	Balochistan
2004	1077	2988	830	217
2005	1239	2500	875	254
2006	1079	3096	1006	284
2007	1066	3293	942	314
2008	961	2912	786	248
2009	1031	3083	921	245
2010	927	3167	986	191
2011	756	2888	953	161
2012	696	2692	1059	163
2013	756	2524	1070	156

Table 3. Fatalities due to terrorism

Year	Civilians	Security force personnel	Terrorists/Insurgents	Total
2004	435	184	244	863
2005	430	81	137	648
2006	608	325	538	1471
2007	1522	597	1479	3598
2008	2155	654	3906	6715
2009	2324	991	8389	11704
2010	1796	469	5170	7435
2011	2738	765	2800	6303
2012	3007	732	2472	6211
2013	3001	676	1702	5379
Total	21202	5474	26837	50327

The data pertaining to terrorist incidents in Pakistan was acquired from South Asian Terrorism Portal [13] and is shown in Table 3.

Global Terrorism data was acquired from "The Global Terrorism Database: Accomplishments and Challenges", a report published by LaFree in 2010 [11].

4 Discussions and Conclusions

The average fatalities caused by road traffic accidents in Pakistan are almost 5067 per year and from terrorist activities 4592 per year. The average fatality rate of traffic accidents as well as terrorist activities is almost same. From the data it is evident that the deaths/fatalities caused by the traffic accidents in Pakistan are almost the same but the fatalities from the terrorist activities is haphazard and does not follow a trend.

Large variation occurs in the data of fatalities caused by the terrorist activities as compared to the fatalities caused by the road accidents. As 863 people were killed in 2004 by terrorist activities and maximum deaths occurred in 2009 are 11704. Due to traffic accidents 4672 deaths occurred in 2013 and maximum deaths of 5615 occurred in 2007. This shows large variation of fatalities caused by terrorist activities and small variation of road traffic fatalities.

The global traffic data of 2013 revealed that least traffic fatalities occurred in Australia numbering to 1192, whereas maximum fatalities occurred in India, a total of 137572.

The graph presented as Fig. 2 has been generated on the basis of data obtained from "World Health Organization-Violence and injury prevention database" [14].

The global terrorism data from 1970–2007 revealed that maximum deaths that occurred due to terrorism activities were in Iraq.

Fatalities caused by terrorist activities in different countries during 1970–2007 [12] have been presented in Fig. 3.

The price for safeguarding human lives against terrorist activities is quite high which includes vast amount of man power, policy changes and monetary cost, the calculation of which is beyond the scope of this paper.

On the other hand the price of safeguarding human lives against traffic fatalities is lower. This can be achieved by slight psychological changes in the mind set of drivers, minor policy changes and low monetary cost.

Figure 1 shows that the number of RTFs is constant through 2004–2013 whereas deaths caused by terrorism have increased significantly. As stated earlier, fatalities due to terrorism are difficult to control due to involvement of numerous as well as arbitrary factors due to which a systematic solution cannot be proposed. However, but by taking some remedial measures we can easily control fatalities occurring due to RTAs because the main cause of traffic accidents is due unawareness of rules and guideline. Therefore, there is a need of Institutes to create awareness about traffic rules. Lax implementation of traffic laws is another cause of traffic accidents.

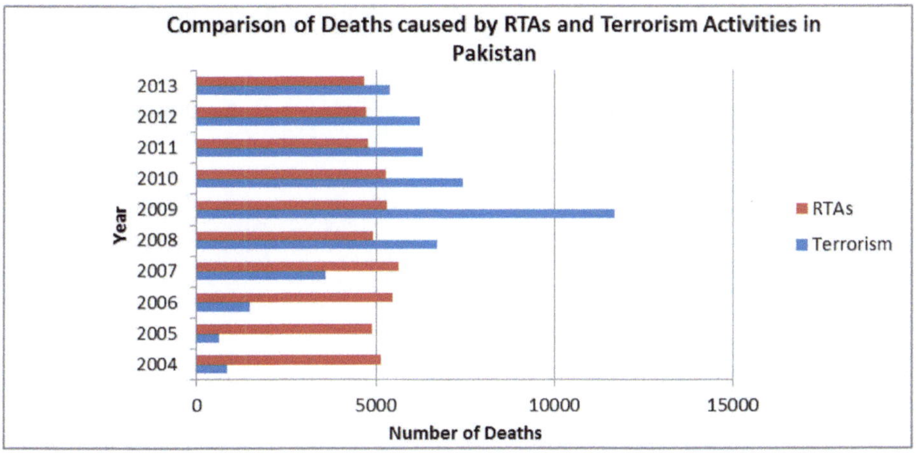

Fig. 1. Comparison of deaths caused by RTAs and terrorism activities in Pakistan

Other causes of road accidents are overloading, drugs addiction by majority of the drivers and use of cell phone during driving. All these problems can be resolved by building awareness and by applying some penalty to defaulters. By the strong system of check and balance by any administrative power we can easily fix these problems. Research has shown that driver error accounts for over 80% of all fatal and injury crashes. Thus by taking certain measures, RTAs can be significantly reduced.

4.1 Improvement Measures

4.1.1 To Strengthen Road Safety Legislation

Road safety laws develop road user behavior to reduce road traffic crashes, injuries and deaths. A number of countries have attained nonstop declines in traffic-related injuries

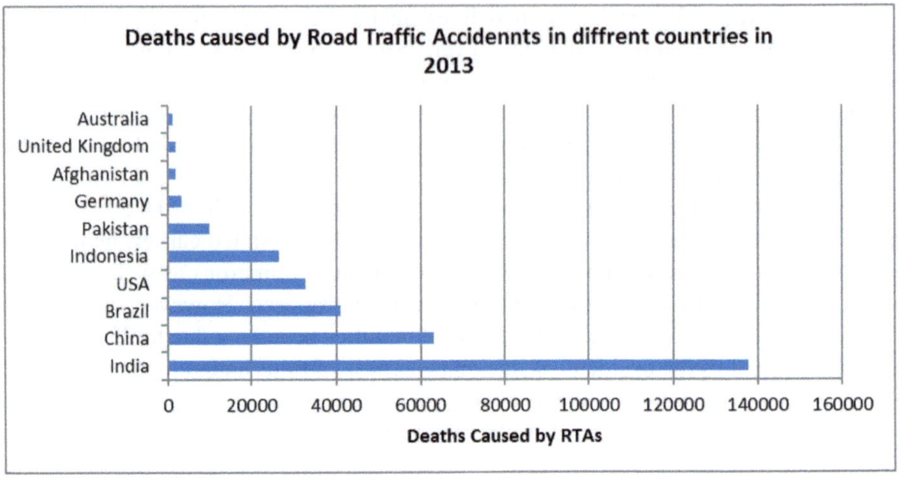

Fig. 2. Deaths caused by road traffic accidents in different countries

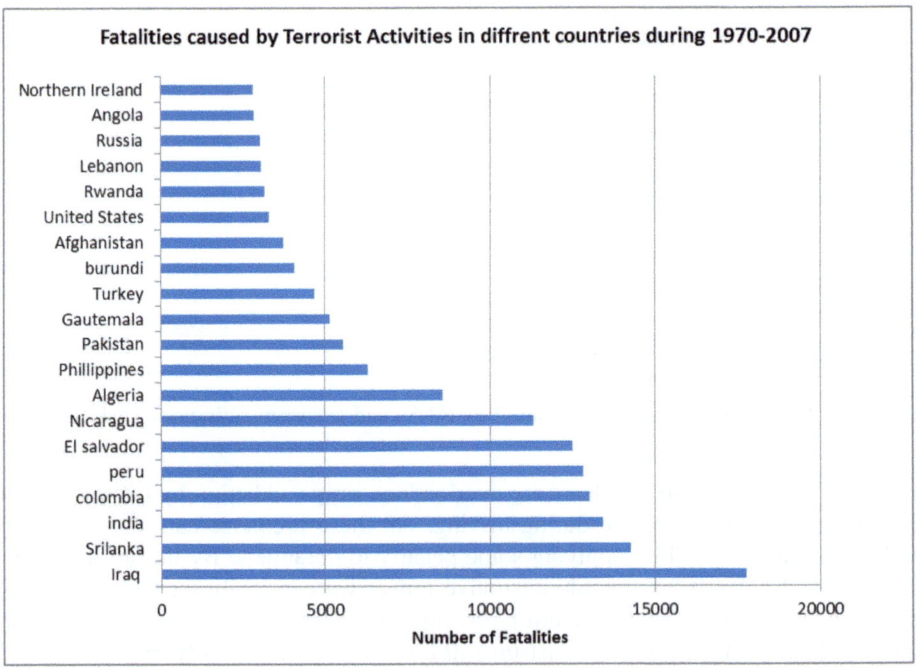

Fig. 3. Fatalities caused by terrorist activities in different countries during 1970–2007 [10]

and fatalities through effective road safety programs that have included legislative change [4]. The most positive changes to road user behavior happen when road safety legislation is reinforced by strong and continuous enforcement, and where the public is

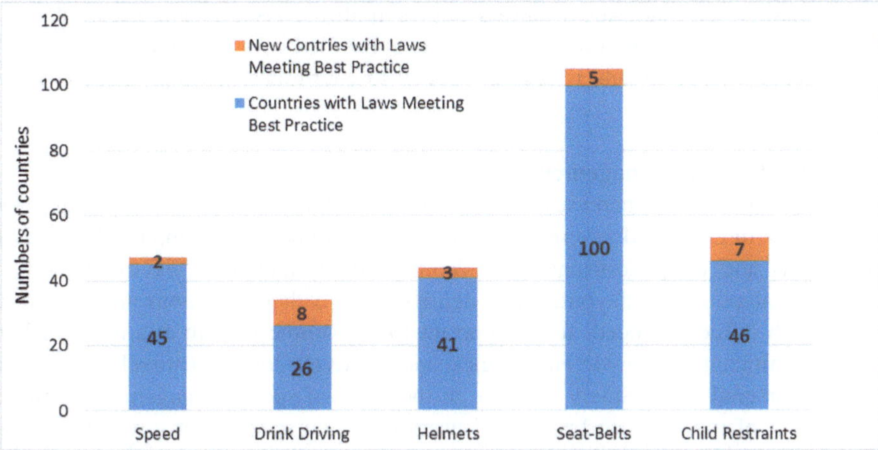

Fig. 4. Changes in legislation on behavioral risk factors 2011–2014 (number of countries represented) [14]

made conscious of the reasons behind the new law and the consequences of nonfulfillment. While there is clear evidence that enforcement is critical to the accomplishment of laws, the levels of enforcement prerequisite for maximum impact are often less readily available and depend on factors such as political will, available resources and competing significances at a national level. Figure 4 shows changes in legislation on behavioral risk factors 2011–2014 (number of countries represented) [15].

4.1.2 Improvements of Emergency Care Facilities

The gross inequalities in injury outcomes between high-income countries and low and middle-income countries narrate directly to the level of care received instantly post-crash and later in a health-care facility. If trauma care systems for seriously injured patients in low and middle-income countries could be brought up to the levels of high performing countries, a predictable half a million lives could be saved each year [15]. There should be single emergency national access number, according to global status report on road safety 2015. Currently, 116 countries have a universal access number to initiate emergency service response.in Pakistan this facility is only available on motorways network.

This compares to 111 countries which had this number in place in 2010. Preferably, a universal emergency telephone number should:

- be usable throughout the catchment area;
- be accessible from every telephone device (landline or mobile);
- be easy to remember and dial (i.e. limited to 3 or 4 digits);
- be free of charge;
- provide access to a nearby vehicle report center;
- assure the confidentially of the caller.

Coordination of road safety efforts across multiple sectors and stakeholders is critical for success. In many countries this role is fulfilled by a lead agency that should ideally have the authority and resources needed to coordinate the implementation of a national strategy [15].

4.1.3 Speed Limit Management

As average traffic speed increases, so does the probability of a crash. If a crash does occur, the risk of death and severe injury is greater at higher speeds mainly for cyclists, motorcyclists and for pedestrians, [15]. Young drivers and male are more likely to speed, while alcohol, road layout, traffic density and weather conditions are some other factors which influence speed. It is important for economic growth at a national level and also significant to provide easy, quick and relatively low-cost travel for people's work and personal lives. Safety must be at the heart of speed management (using an integrated set of measures to bring road users to a safe speed), yet governments and those involved in speed management at local level face challenges when harmonizing mobility and safety. However, shifting the stress towards safety is at the heart of the "Safe System" approach, The Safe System approach which is for accommodating human error and approach to road safety ensures that, in a crash, impact energy remains below the verge likely to result in death or severe injury. It goes beyond establishing speed limits to managing connections between the environment, infrastructure and physical exposure.

Within this approach, speed limits are a complementary interference to creating safer roads, vehicles and roadsides and that work together for accommodating driver error. All parts of the system need to be reinforced – roads, roadsides, speed limits and vehicles – so that if one part of the system fails, other parts will still guard people involved [15].

Speed limit must be under 30 km/h where pedestrians and cyclists mix with the motorized traffic, because at high speed these road users are more vulnerable to an accident. An adult pedestrian has less than 20% chance of dying if hit by a car at less than 50 km/h but if hit at 80 km/h has almost a 60% risk of dying [7]. An important step in reducing speed is setting and imposing a national speed limit. Most of the countries has general national speed limits of restricted number, for example for urban, motorways, and rural roads, and some countries further divide it (such as between "urban industrial" and "urban residential" areas). In 97 countries out of 180 contributing countries in the developing road safety, a speed limits of less than or equal to 50 km/h is the maximum speed for urban roads [14]. Also the definition of urban vary between countries, a speeds above 50 km/h would be unsafe as these areas usually have a high concentration of cyclists and pedestrians.

4.1.4 Safety Measure for Motorcyclist

Avoiding motorcyclist head injuries is becoming increasingly crucial as motorcycle use increases. According to global report on road safety 2015 shows that between 2010 and 2013 there was a 27% growth in the number of motorized two-wheelers globally. Motorcycles form a high proportion of vehicle fleets in Pakistan as well that is app: 55% of total registered vehicles.

Motorcyclists are at a bigger risk because they frequently share the traffic space with fast-moving cars, buses and trucks, and because they are less noticeable.

Head injuries also effect in much higher medical costs than any other type of injury. Wearing a motorcycle helmet can decrease the risk of death by nearly 40% and the risk of severe injury by approximately 70% [15]. Active implementation of motorcycle helmet laws can increase helmet-wearing rates and thereby reduce head injuries.

Figure 5 indicates the status of motor vehicles registered in Pakistan till 2015.

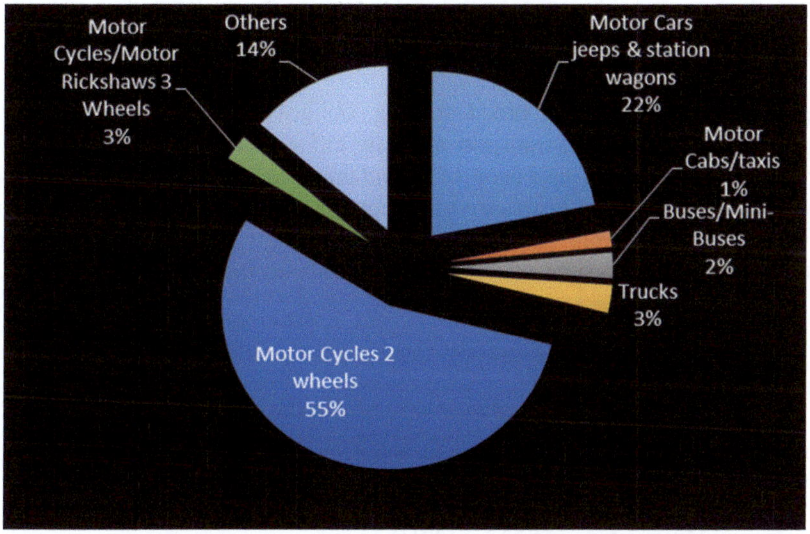

Fig. 5. State of motor vehicles registered in Pakistan 2010

4.1.5 Increasing Seat-Belt Use

In the occasion of a crash seat-belts limit the movement of vehicle occupants, scattering the force of the restraint to decrease the probability of severe or fatal injury.

Wearing a seat-belt decreases the hazard of a casualty for front-seat occupants and drivers by 45–50% and the risk of serious and minor injuries by 45% and 20% respectively. Among rear-seat occupants seat-belts reduce minor injuries by up to 75% and fatal and serious injuries by 25% [15].

4.1.6 Making Roads Safer

At the cost of safety, mobility and economic efficiency has traditionally maximized due to road infrastructure, mainly for non-motorized road users who are the utmost susceptible.

Indeed, walking and cycling have become less common and more unsafe in many countries as motorization rises worldwide, The traffic mix in many countries means that pedestrians and cyclists share the road with high-speed vehicles, compelling them to exchange dangerous circumstances and fast-moving traffic. Planning decisions has to be made without adequate attention to the requirements of these groups – for example, cycle paths and footpaths are often not part of an integrated network. At the same time, traffic congestion resulting from rapid motorization means the transport and mobility demands of local publics are frequently not met. Modifications are now obligatory which are now moving quickly towards much advanced levels of motorization, improved levels of air pollution and more sedentary lifestyles.

A key policy for attaining a safe traffic system for pedestrians and cyclists is to separate these different kinds of road use, removing conflicts between high-speed and susceptible road users. Safety benefits of measures such as building separate cycle lanes [15].

5 Recommendations

Addressing road safety in a broad manner obliges a coordinated multi-sectorial approach connecting health, transport and law enforcement sectors. Construction of well-planned road system; provision of ambulances and pre-hospital care and emergency medical services are necessary to reduce post-accident fatalities. It is also very important to increase public awareness about road safety. Drivers need to be trained and educated to give the right of way to pedestrians on the roads.

References

1. United Nations: Department of Economic and Social Affairs. World Population Prospects: The 2015 revision (2015)
2. WHO: Road Traffic Injuries. From WHO (2016). http://www.who.int/mediacentre/factsheets/fs358/en/. Accessed 25 July 2016
3. Statista: The Statisitcs Portal (n.d.). From Statista (2016). http://www.statista.com/statistics/202871/number-of-fatalities-by-terrorist-attacks-worldwide/. Accessed 26 July 2016
4. Fatmi, Z., et al.: Incidence, patterns and severity of reported unintentional injuries in Pakistan for persons five years and older: results of the National Health Survey of Pakistan 1990–1994. BMC Public Health **7**(1), 152 (2007)
5. Bishail, D., Hyder, A.A., Ghaffar, A., Morrow, R.H., Kobusingye, O.: Rates of public investment for road safety in developing countries: case studies of Uganda and Pakistan. Health Policy Plann. **18**(2), 232–235 (2003)
6. Luby, S., et al.: Road traffic injuries in Karachi: the disproportionate role of buses and trucks. SE Asian J. Trop. Med. Public Health **28**, 395–398 (1997)
7. Shamim, S., et al.: Initial results of Pakistan's first road traffic injury surveillance project. Int. J. Inj. Control Saf. Promot. **18**(3), 213–217 (2011)
8. Ghani, E., et al.: Road traffic accidents as a major contributor to neurosurgical mortality in adults. J. Coll. Phys. Surg.–Pakistan: JCPSP **13**(3), 143–145 (2003)
9. Naci, H., Chisholm, D., Baker, T.D.: Distribution of road traffic deaths by road user group: a global comparison. J. Inj. Prev. **15**, 55 (2009)
10. Reports on International Terrorism: Statistics on Incidents of Terrorism Worldwide. http://www.jewishvirtuallibrary.org/jsource/Terrorism/terrorstats.html
11. LaFree, G.: The global terrorism database: accomplishments and challenges, Prespect. Terrorism **4**(1) (2010)

12. Statistics, P.B.: Pakistan Bureau of Statistics (n.d.). From Pakistan Bureau of Statistics: http://www.pbs.gov.pk/sites/default/files/social_statistics/crime_statistics/traffic_accidents.pdf. Accessed 25 July 2016
13. Portal, S.A.: Fatalities in Terrorist Violence in Pakistan 2003–2016. South Asia Terrorism Portal (24 July 2016). http://www.satp.org/satporgtp/countries/pakistan/database/casualties.htm. Accessed 26 July 2016
14. WHO: Violence and Injury Prevention. From WHO: 13 (2015). http://www.who.int/violence_injury_prevention/road_safety_status/2015/GSRRS2015_data/en/. Accessed 27 July 2016
15. Global Status Report on Road Safety 2015

Author Index

J. Calautit et al. (eds.), *Towards Sustainable Cities in Asia and the Middle East*,
Sustainable Civil Infrastructures, DOI 10.1007/978-3-319-61645-2

MIX
Papier aus verantwortungsvollen Quellen
Paper from responsible sources
FSC® C105338

If you have any concerns about our products,
you can contact us on
ProductSafety@springernature.com

In case Publisher is established outside the EU,
the EU authorized representative is:
Springer Nature Customer Service Center GmbH
Europaplatz 3, 69115 Heidelberg, Germany

Printed by Libri Plureos GmbH
in Hamburg, Germany